STUDY GUIDE

Discover Biology

Core Topics

Fifth Edition

Anu Singh-Cundy • Michael Cain

D1377122

Discover Biology

Core Topics

Fifth Edition

Anu Singh-Cundy • Michael Cain

Neva Laurie-Berry

PACIFIC LUTHERAN UNIVERSITY

Margaret Liberti

SUNY COBLESKILL

 W • W • NORTON & COMPANY • NEW YORK • LONDON

W. W. Norton & Company has been independent since its founding in 1923, when William Warder Norton and Mary D. Herter Norton first published lectures delivered at the People's Institute, the adult education division of New York City's Cooper Union. The firm soon expanded its program beyond the Institute, publishing books by celebrated academics from America and abroad. By midcentury, the two major pillars of Norton's publishing program—trade books and college texts—were firmly established. In the 1950s, the Norton family transferred control of the company to its employees, and today—with a staff of four hundred and a comparable number of trade, college, and professional titles published each year—W. W. Norton & Company stands as the largest and oldest publishing house owned wholly by its employees.

Associate editor, ancillaries: Callinda Taylor
Production manager: Eric Pier-Hocking
Composition by Westchester Book Group
Manufacturing by Sterling Pierce

ISBN 978-0-393-91854-0

W. W. Norton & Company, Inc.
500 Fifth Avenue, New York, N.Y. 10110-0017

wwnorton.com

W. W. Norton & Company Ltd.
Castle House, 75/76 Wells Street, London W1T 3QT

4 5 6 7 8 9 0

TABLE OF CONTENTS

Unit 1 The Diversity of Life

Unit 2 Cells: The Basic Units of Life

Unit 3 Genetics

Unit 4   Evolution

Unit 5   Ecology

CHAPTER 1 | The Nature of Science and the Characteristics of Life

GETTING STARTED

Below are a few questions to consider before reading Chapter 1. These questions will help guide your exploration and assist you in identifying the key concepts presented in this chapter.

1. What is the scientific method, and why do scientists use it to study the natural world?

2. What are the characteristics that all living things have in common?

3. What is biological evolution, and what can cause genetic characteristics to change over time?

4. What is a biological hierarchy, and how does knowing about it help us in our everyday lives?

5. Two types of reproduction are associated with multicellular living organisms. Identify these types of reproduction, and identify the genetic material used to perform reproduction.

6. What is the difference between an ecosystem and a biome?

7. What are microbes, and are bacteria included in this classification?

A GUIDE TO THE READING

The following concepts typically give students the most difficulty when exploring the content in Chapter 1 for the first time. For each concept, one or more references have been identified that may help you gain a better understanding of these potentially problematic areas.

Theory vs. Fact

In scientific terms, a fact is a repeatable observation, whereas a theory is a testable, plausible explanation for a natural phenomenon. The term "theory" may, however, be misunderstood as nothing more than a well-educated guess to explain a given observation. It is important to understand that a scientific theory is actually an idea or explanation that has withstood rigorous investigation through years of experimentation. As such, a theory is much more than an educated guess and, in most cases, provides the most likely explanation for the natural phenomenon or observation in question.

For more information on this concept, be sure to focus on

- Section 1.1, *A scientific theory is a body of knowledge that has stood the test of time*

What Is Life?

Characteristics that are common to all living organisms help us better define what makes something "alive." Seven properties are described in Chapter 1 that are common to all living things. (Recall that all living organisms [1] are composed of cells, [2] reproduce using the hereditary material deoxyribonucleic acid [DNA], [3] grow and develop, [4] capture energy from their environment, [5] sense their environment and respond to it, [6] maintain a constant internal environment, and [7] evolve.) Consider examples of how we possess each of these characteristics.

For more information on this concept, be sure to focus on

- Section 1.3, The Characteristics of Living Organisms

Evolution vs. Adaptation

A group of organisms capable of breeding with one another and producing viable offspring is referred to as a "species." Within a population, each new generation of a species has the potential to introduce genetic change. If these changes impart a survival or reproductive advantage to individuals possessing any new characteristics caused by the change or changes, the changes can be considered adaptations. Evolution occurs when adaptations accumulate within a population over time and alter one or more characteristics that define the species.

For more information on this concept, be sure to focus on

• Section 1.3, *Groups of living organisms can evolve*

TYING IT ALL TOGETHER

Not surprisingly, concepts presented in this introductory chapter will be revisited and discussed in greater detail in subsequent chapters, including

Cell Theory of Life

• Chapter 6—Section 6.1, Cells: The Smallest Units of Life

Energy Flow in a Living System

• Chapter 8—Section 8.1, The Role of Energy in Living Systems
• Chapter 8—Section 8.2, Metabolism

DNA as Hereditary Material

• Chapter 14—Section 14.1, An Overview of DNA and Genes

Evolution

• Chapter 17—Section 17.1, Descent with Modification
• Chapter 17—Section 17.5, The Impact of Evolutionary Thought
• Chapter 20—Section 20.2, The History of Life on Earth

Responding to the Environment

• Chapter 30—Section 30.4, Sensory Structures: Making Sense of the Environment

Reproduction and Development

• Chapter 33—Section 33.1, Sexual and Asexual Reproduction in Animals

The Biological Hierarchy and Food Webs

• Chapter 21—Section 21.3, Terrestrial Biomes
• Chapter 21—Section 21.4, Aquatic Biomes
• Chapter 23—Section 23.2, How Species Interactions Shape Communities

PRACTICE QUESTIONS

Factual Knowledge

1. Which of the following is an example or type of evolution?
   a. artificial selection
   b. biological evolution
   c. natural selection
   d. domestication
   e. all of the above

2. A newly discovered microscopic structure is hypothesized to be a living organism. Which of the following characteristics would support the contention that this "organism" may be alive?
   a. It contains DNA.
   b. It consists of a single cell.
   c. It absorbs energy.
   d. It maintains a constant internal environment.
   e. all of the above

3. Consider the brain labeled in Figure 1.12 of your textbook. In terms of the biological hierarchy described in the chapter, which of the following structures are contained within this body part?
   a. molecules
   b. cells
   c. organ systems
   d. tissues
   e. More than one of the above is correct.

4. Advancing biological knowledge depends on rigorously testing hypotheses about the living world. (True or False)

5. Scientists begin constructing a simplified view of nature by doing experiments and forming theories. (True or False)

6. Which of the following would you expect to find in common among cells from a rose, a mushroom, and a cat?
   a. DNA
   b. tissues
   c. viruses
   d. organs
   e. biospheres

7. Evolution represents biological change that accumulates over time. However, this change is not always apparent at every level in the biological hierarchy. Scientists most commonly observe the effects of evolution at the level of the
   a. species.
   b. community.
   c. cell.
   d. organism.
   e. ecosystem.

8. A prediction is made before observations or hypotheses have been formed. (True or False)

9. If a cell contains DNA and other biological molecules, maintains a constant internal environment, and shows signs of having once grown and developed, then it must currently be alive. (True or False)

10. Consider the studies conducted by Dr. Burkholder to determine the cause of the mysterious die-offs plaguing fish in the rivers of North Carolina described in the chapter. Her postulation that the deaths were caused by a type of organism called a "protist" is considered a(n)
    a. experiment.
    b. observation.
    c. hypothesis.
    d. theory.
    e. test.

11. Match each term with the best description.
    ___ adaptation
    ___ ecosystem
    ___ evolution
    ___ homeostasis
    ___ hypothesis
    ___ species
    a. process of maintaining a constant internal environment
    b. a collection of organisms and the places where they live
    c. biological change accumulating in a species over time
    d. a group of interbreeding individuals capable of producing fertile offspring
    e. features that provide a survival or reproductive benefit
    f. tentative, testable explanation for an observation

12. Which of the following is *not* a potential source of energy for a food web?
    a. plants
    b. animals
    c. the sun
    d. nonliving materials, such as iron
    e. All of the above are potential sources of energy.

Conceptual Understanding

1. Which of the following is *not* a scientifically testable hypothesis?
   a. The growth of peanut plants is affected by soil nitrogen concentration.
   b. El Niño–induced weather patterns influence fish harvests in the Pacific.
   c. Biological diversity is the result of a single divine creation.
   d. High dietary fiber leads to lower cholesterol in humans.
   e. Natural fires lead to the rejuvenation of certain plant communities.

2. Your television won't turn on, and you speculate that a circuit breaker in the house has been tripped. In scientific terminology, such reasoning would best be described as
   a. forming conclusions from the results of experiments.
   b. controlling variables in a repeated manipulation of nature.
   c. developing an observation based on a hypothesis.
   d. developing a hypothesis based on an observation.
   e. testing a prediction generated from a hypothesis.

3. Consider Burkholder's experiments on the mysterious die-offs of fish in North Carolina rivers. Which of the following best represents her main hypothesis?
   a. Fish maintained in laboratory environments were dying when exposed to river water.
   b. Microscopic organisms called "protists" accumulated in large numbers in laboratory fish tanks just before the fish died.
   c. Microscopic organisms called "protists" were responsible for causing the fish die-offs.
   d. Fish maintained in laboratory environments and not exposed to river water did not die.
   e. none of the above

4. In the article "Earthbound Extraterrestrial? Or Just Another Microbe in the Mud?" at the beginning of Chapter 1 and "Researchers Wrangle over Bacteria" at the end of Chapter 1, the phenomenon of microbes growing in strong concentrations of arsenic is considered unique. Why?
   a. All organisms reproduce themselves using DNA, and arsenic kills DNA.
   b. All organisms require DNA in order to reproduce, and arsenic removes phosphorous from DNA strands.
   c. All organisms reproduce themselves using DNA, and arsenic's similar structure replaces phosphorous.

d. Arsenic removes and replaces phosphorous on DNA and doesn't allow DNA to replicate.

e. none of the above

5. Characteristics that are beneficial to species, such as the average speed at which pronghorn antelope can run, are known as evolutionary _____ and are determined largely by the _____ of an individual.

a. changes; DNA
b. adaptations; DNA
c. dead ends; DNA
d. hypotheses; energy
e. food webs; energy

6. In most living systems, the energy that powers life comes ultimately from

a. plants.
b. animals.
c. fungi.
d. bacteria.
e. the sun.

7. An important feature of most food webs is that energy

a. ultimately comes from sources other than the sun.
b. flows in one direction, from animals to plants and then to the sun.
c. flows in one direction, from the sun to producers and then to consumers.
d. flows in multiple directions, from plants to animals and then back again.
e. seems to follow no particular pattern.

8. In a particular food web, the leaves of oak trees are eaten by moth caterpillars. Some of the caterpillars are then eaten by birds, which in turn are eaten by snakes. In this food web, the consumers are the

a. oak trees.
b. oak trees and caterpillars.
c. caterpillars and birds.
d. birds and snakes.
e. caterpillars, birds, and snakes.

9. Using the article "Earthbound Extraterrestrial? Or Just Another Microbe in the Mud?" determine which of the following would be a feasible hypothesis.

a. If microbes do not have phosphorous, they will die.
b. If microbes can live in toxic arsenic, they are extraterrestrial.
c. If microbes can live in toxic arsenic, they have evolved.
d. If microbes are extraterrestrial, they will survive harsh conditions.
e. If microbes can live in toxic arsenic, any organism can change its DNA composition.

10. Consider the following list of characteristics common to all living organisms. Match the characteristics with the appropriate example.

___ reproduction using DNA
___ actively taking in energy from the environment
___ sensing and responding to the environment
___ maintaining a constant internal environment
___ evolving

a. Sweating occurs as a result of vigorous exercise.
b. Pronghorns with the ability to run faster have a survival advantage over slower pronghorns.
c. Some plants have the ability to undergo photosynthesis as well as capture and devour insects.
d. Characteristics are passed from parents to offspring.
e. Sunflowers typically turn toward the warmth of the sun.

11. In which level of the biological hierarchy (Figure 1.12) would the microbes from the articles at the beginning and end of the chapter have been changed because of evolution with arsenic?

a. atoms
b. molecule (DNA)
c. cell
d. tissue
e. organ

12. The microbes have exhibited which type of evolution?

a. adaptation
b. natural selection
c. artificial selection
d. biological evolution
e. none of the above

RELATED ACTIVITIES

• Observe Figure 1.12, The Biological Hierarchy Extends from the Atom to the Biosphere. Pick a specific organism in a population and determine how that organism would fit into the hierarchy with its appropriate components. Then determine how changing the gene expression of that organism would affect the entire hierarchy. After you have completed this task, use your new understanding of the biological hierarchy to determine how a disease or predator that would wipe out an entire population would affect this particular hierarchy.

• Review the Biology Matters box, "Science and the Citizen," in Chapter 1. This passage describes the effect of science on the U.S. citizen. Science articles and phenomena exist that are controversial and regularly cause great debate. Using at least two outside sources, find two science discoveries that will affect U.S. citizens now or in the near future, and describe how you are going to handle this situation. Determine what you would do if you were (1) the scientist who discovered this controversial phenomenon and (2) one of the politicians who was fighting this discovery.

- We live in a society in which it is no longer reasonable to believe that people can remain scientifically illiterate and still function well in the modern world. Review Section 1.4, The Biological Hierarchy. At the end of this section are several examples of how understanding this hierarchy is important for everyday life. Review each of the examples and conduct the following exercise: Using your local newspaper, a magazine, or the Internet, identify an article that demonstrates humankind's interaction with the natural world at each of the levels of the hierarchy. Compose a one-page summary of your examples to show how understanding our relationship with the natural world can be beneficial to our everyday lives.

ANSWERS AND EXPLANATIONS

Factual Knowledge

1. e. All of the above. Domestication is a type of artificial selection, and both constitute adaptive evolution by humans. Biological evolution is evolution of genetic characteristics, and natural selection is "survival of the fittest," the most typical explanation of evolution students know. For more information, see Section 1.3, *Groups of living organisms can evolve.*

2. e. All of these characteristics are shared by all living organisms. However, to be considered "living," an organism must possess all of the characteristics described in the chapter. For more information, see Section 1.3, The Characteristics of Living Organisms.

3. e. Recall that the brain is considered an organ. In terms of the biological hierarchy, organs are built from a collection of tissues that contain cells assembled from molecules. The brain consequently would be considered part of the nervous system, which controls our ability to interact with the environment. For more information, see Figure 1.12 and Section 1.4, The Biological Hierarchy.

4. True. Rigorous hypothesis testing is the core of scientific inquiry. For more information, see Section 1.1, The Nature of Science and *Science is a body of knowledge and a process for generating that knowledge.*

5. False. The first steps in the scientific method are observation and hypothesis formation. Experiments are subsequently performed to test the hypothesis. The experimental results are then analyzed and used to support or reject the hypothesis. The term "theory" is reserved for well-tested and well-supported hypotheses.

For more information, see Section 1.1, *A scientific theory is a body of knowledge that has stood the test of time.* See also Section 1.1, *Experiments are the gold standard for establishing causality.*

6. a. DNA is universal in all cells. Although viruses may often infect cells, typically they are not present in healthy cells. The other choices are all structures larger than a cell. For more information, see Section 1.3, *Living organisms reproduce themselves via DNA.* There is also some background information under Section 1.3, *Living organisms are composed of cells.*

7. a. Evolution is first evident at the species level, mostly for reasons having to do with the inheritance of adaptations within a population. For more information, see Section 1.3, *Groups of living organisms can evolve.*

8. False. Predictions are made *after* observations have been carried out and a hypothesis has been formed. For more information, see Figure 1.2 and Section 1.1, *Science is a body of knowledge and a process for generating that knowledge.*

9. False. Consider that a dead cell would also show all of the characteristics listed. For a cell to be considered alive, it must possess *all* of the characteristics described in the chapter, including energy utilization and the ability to sense and respond to its environment. For more information, see Section 1.3, The Characteristics of Living Organisms.

10. c. Recall that a hypothesis is a testable explanation for a given observation. In this case a test, or experiment, was used to determine the validity of the proposed hypothesis. For more information, see Section 1.1, *Scientific hypotheses must be testable and falsifiable.*

11. e. adaptation. For more information, see Section 1.3, *Groups of living organisms can evolve.*
    b. ecosystem. For more information, see Section 1.4, The Biological Hierarchy.
    c. evolution. For more information, see Section 1.3, *Groups of living organisms can evolve.*
    a. homeostasis. For more information, see Section 1.3, *Living organisms actively maintain their internal conditions.*
    f. hypothesis. For more information, see Section 1.1, *Scientific hypotheses must be testable and falsifiable.*
    d. species. For more information, see Section 1.3, *Groups of living organisms can evolve.*

12. e. Recall that most food webs begin with the capture of sunlight energy by organisms that fall into the producer category, including most plants. Consumers, including most animals, may then prey upon both producers and other consumers for their own energy needs. Some biological systems devoid of light (such as the bottom of the ocean) may rely on energy derived from nonliving material for sustenance. For

more information, see Section 1.3, *Living organisms obtain energy from their environment.*

Conceptual Understanding

1. c. Using divine creation as an explanation for biodiversity requires a faith-based hypothesis that an all-powerful deity exists. Science cannot work with such hypotheses because they are inherently untestable. For more information, see Section 1.1, *Scientific hypotheses must be testable and falsifiable.*

2. d. This is the process of going from observation (the television won't turn on) to tentative explanation (a hypothesis about the circuit breaker). This is a synopsis of all of Section 1.1, but for more information on hypotheses, see Section 1.1, *Scientific hypotheses must be testable and falsifiable.* For more information on observations, see Section 1.1, *Observations are the wellspring of science.*

3. c. This is the only choice that provides a possible and testable explanation for the fish die-offs. The other choices are simply observations about the phenomenon. For more information on hypotheses, see Section 1.1, *Scientific hypotheses must be testable and falsifiable.* See also the discussion of Burkholder's experiments in Section 1.1.

4. c. Arsenic has a similar structure to phosphorous, which is on the backbone of the DNA strand. Arsenic typically is toxic to DNA because it depletes the phosphorous in a cell (which is necessary to make new DNA and therefore new cells). This would eventually cause the death of cells, except that over time these microbes have adapted the arsenic into their DNA in place of phosphorous, allowing them to survive in toxic amounts of arsenic. (The microbes are a type of extremophile.) In Chapter 1, the articles "Earthbound Extraterrestrial? Or Just Another Microbe in the Mud?" and "Researchers Wrangle over Bacteria" will help answer this question.

5. b. Evolution is caused by the accumulation of change within a species over time, and adaptations are the changes that provide survival or reproductive benefits. The majority of adaptations are products of changes in an organism's DNA, which can be passed on to subsequent generations. For more information on evolution, see Section 1.3, *Groups of living organisms can evolve.* For more information on DNA, see Section 1.3, *Living organisms reproduce themselves via DNA.*

6. e. Most living systems ultimately derive their energy from the sun. However, there are examples, such as ecosystems present at the bottom of the ocean, that do not rely ultimately on the sun for energy. For more information, see Section 1.3, *Living organisms obtain energy from their environment.*

7. c. Refer to Figure 1.8, and note how the energy flows from the sun to producers and then to consumers of one kind or another. For more information, see Section 1.3, *Living organisms obtain energy from their environment.*

8. e. The oak trees (plants) are the producers; the remaining organisms in the scenario would be considered a type of consumer. For more information, see Section 1.3, *Living organisms obtain energy from their environment.*

9. c. The microbes have evolved to allow arsenic, which structurally resembles phosphorous, to replace phosphorous in the backbone of the DNA strand. Normally, arsenic would be toxic and kill bacteria (microbes). However, these microbes have evolved and allowed arsenic to be incorporated. For more information, see Section 1.1, *Scientific hypotheses must be testable and falsifiable.* See also Chapter 1, the article "Earthbound Extraterrestrial? Or Just Another Microbe in the Mud?" and "Researchers Wrangle over Bacteria."

10. d. reproduction using DNA. For more information, see Section 1.3, *Living organisms reproduce themselves via DNA.*
    c. actively taking in energy from the environment. For more information, see Section 1.3, *Living organisms obtain energy from their environment.*
    e. sensing and responding to the environment. For more information, see Section 1.3, *Living organisms sense their environment and respond to it.*
    a. maintaining a constant internal environment. For more information, see Section 1.3, *Living organisms actively maintain their internal conditions.*
    b. evolving. For more information, see Section 1.3, *Groups of living organisms can evolve.*

11. b. For more information, see Figure 1.12; the articles "Earthbound Extraterrestrial? Or Just Another Microbe in the Mud?" and "Researchers Wrangle over Bacteria"; Section 1.4, The Biological Hierarchy; and Section 1.3, *Living organisms reproduce themselves via DNA.*

12. b. Natural selection occurs when organisms survive some type of external stimuli or predator (such as arsenic) and those that survived reproduce. The bacteria that survived the arsenic exposure and took the arsenic into their genetic makeup or backbone were the organisms to survive and reproduce, passing on those survival genes. For more information, see Section 1.3, *Groups of living organisms can evolve.*

CHAPTER 2 | Biological Diversity, Bacteria, and Archaea

GETTING STARTED

Below are a few questions to consider before reading Chapter 2. These questions will help guide your exploration and assist you in identifying the key concepts presented in this chapter.

1. It is believed that all life on earth was derived from a single common ancestor. How do scientists determine the ways in which groups of organisms are related to one another?

2. Why is an evolutionary tree considered to be a scientific hypothesis?

3. What are shared derived features, and how do they differ from convergent features when determining evolutionary relationships?

4. How is the "wise man" (*Homo sapiens*) related to the "upright man" (*Homo erectus*) and the "handy man" (*Homo habilis*)?

5. Why is the mushroom on your slice of pizza more closely related to you than it is to the green pepper that sits next to it?

6. Why have bacteria and archaea been successful in surviving and growing on earth over long periods?

7. Are viruses considered alive or dead? How do viruses have the ability to cause so much sickness?

A GUIDE TO THE READING

The following concepts typically give students the most difficulty when exploring the content in Chapter 2 for the first time. For each concept, one or more references have been identified that may help you gain a better understanding of these potentially problematic areas.

Common Ancestor

Just like a family tree, an evolutionary tree depicts relationships—in this case, relationships among different species or groups of organisms. Descendants from a single ancestral species are depicted on the evolutionary tree as the tip of a branch. The most recent common ancestor of two divergent species is therefore represented on the tree by the point where one lineage branch joins another. Although two groups of organisms may have many common ancestors, there can be only one most recent common ancestor, represented by the fork on the tree.

For more information on this concept, be sure to focus on

- Section 2.1, *A common origin explains the unity of life on Earth*
- Section 2.1, *Evolutionary divergence explains the diversity of life on Earth*
- Figure 2.2, Evolutionary Tree of Domains

Domains

In the chapter, the Linnaean hierarchy is described as a system adopted by scientists to classify organisms into specific taxonomic groups reflecting their relationships with other organisms within the hierarchy. The original hierarchy described just two kingdoms: plants and animals. Over time, scientists have expanded this system to include upwards of 13 kingdoms of organisms (our textbook adopts the widely used six-kingdom system). Beyond this, scientists have recognized a need to create even larger groupings of organisms, referred to as "domains." There are three proposed domains: Bacteria, Archaea, and Eukarya. When reviewing the chapter, be sure to pay particular attention to the key differences that distinguish the Bacteria from the Archaea and to how scientists have been able to use DNA to help support this proposed domain structure.

For more information on this concept, be sure to focus on

- Section 2.1, *All of life on Earth can be sorted into three distinct domains*
- Section 2.2, The Linnaean System of Biological Classification
- Figure 2.2, Evolutionary Tree of Domains

Horizontal Gene Transfer

The study of DNA has raised new questions. The text describes an example in which studies revealed the presence of bacterial DNA within the cells of organisms from both the Archaea and the Eukarya domains. Dr. W. Ford Doolittle's theory of horizontal gene transfer was developed to explain how organisms that diverged so long ago could share the same bacterial DNA. In this theory, it is argued that genes (DNA) may actually be passed between different lineages, facilitating the sharing of genes between organisms from different domains.

For more information on this concept, be sure to focus on

- Section 2.1, *All of life on Earth can be sorted into three distinct domains*
- Section 2.3, *Prokaryotes reproduce asexually*
- Figure 2.9, Lateral Gene Transfer Accelerates the Rate of Evolution in Prokaryotes

Multiple Ways to Obtain Nutrients

Life on earth is remarkably adaptable. Organisms exhibit a multitude of ways to obtain nutrients from the environment. Chapter 2 introduces several terms that describe these different methods, particularly in relation to prokaryotes. For a summary of the differences among these organisms, refer to the table at the bottom of this page.

For more information on this concept, be sure to focus on

- Section 2.3, *Prokaryotes are unrivaled in metabolic diversity*

Viruses

Unlike organisms in the domains Archaea, Eukarya, and Bacteria, viruses are not made up of cells. They are not considered alive, but they do have some of the characteristics that allow an organism to be considered alive, as we found in Chapter 1. These include having DNA, the ability to reproduce, and the capacity to evolve. Although viruses do share some similar characteristics with organisms considered alive, they will invade live organisms and take over their cells. A virus uses the host cell to replicate its DNA or ribonucleic acid (RNA) and to either take over that cell or destroy it. Essentially, a virus is genetic material enveloped in a protein coat. Viruses are classified by their shapes similarly to the Linnaean hierarchy discussed earlier in Chapter 2. Retroviruses, such as HIV, can infect an organism but not have an effect on its life for many years.

For more information on this concept, be sure to focus on

- Section 2.4, *Viruses are classified according to structure and type of infection*
- Section 2.4, *Viruses lack cellular organization*
- Figure 2.14, Viruses Can Be Classified by Their Shape
- Figure 2.15, Viruses Reproduce Inside Their Host Cells

TYING IT ALL TOGETHER

Several concepts presented in this chapter will be revisited and discussed in greater detail in subsequent chapters, including

Kingdoms of Life

- Chapters 3 and 4, which review Protista, Plantae, Fungi, and Animalia in more detail

Modern DNA Technologies

- Chapter 14—Section 14.1, An Overview of DNA and Genes
- Chapter 16—Section 16.2, DNA Fingerprinting
- Chapter 16—Section 16.3, Reproductive Cloning of Animals
- Chapter 16—Section 16.4, Genetic Engineering
- Chapter 16—Section 16.5, Human Gene Therapy

Term	Source of Carbon	Energy	Capable of Producing Nutrients on Their Own?
Chemoheterotrophs	Carbohydrates	Chemical	No—From other organisms
Chemoautotrophs	Carbon dioxide	Chemical	Yes
Photoheterotrophs	Carbohydrates	Sunlight	No—From other organisms and carbon-containing compounds
Photoautotrophs	Carbon dioxide	Sunlight	Yes

Evolution and the Origin of Species

- Chapter 17—Section 17.1, Descent with Modification
- Chapter 17—Section 17.4, The Evidence for Biological Evolution
- Chapter 20—Section 20.2, The History of Life on Earth

Humankind's Closest Relatives

- Chapter 20—Section 20.7, Human Evolution

PRACTICE QUESTIONS

Factual Knowledge

1. Most biologists currently estimate the total number of species that exist on Earth to be somewhere between _____ million.
   a. 1 and 2
   b. 2 and 3
   c. 3 and 30
   d. 30 and 100
   e. 100 and 500

2. Which of the following is *not* found in the kingdom Bacteria?
   a. membrane-bound organelles
   b. DNA
   c. organized cell structure
   d. ability to reproduce
   e. ability to use energy

3. A number of reasons exist for the evolutionary success of the kingdoms Bacteria and Archaea. Which of the following characteristics allow these prokaryotes to be so successful?
   a. rapid reproduction
   b. diverse methods of obtaining nutrients
   c. the ability to survive extreme conditions
   d. the ability of some to live with oxygen and others to live without oxygen
   e. all of the above

4. Organisms that share physical traits are almost always members of the same species. (True or False)

5. Prokaryotes derive their energy from a variety of sources. Which of the following terms does *not* accurately describe prokaryotes?
   a. chemoheterotrophs
   b. chemoautotrophs
   c. photoheterotrophs
   d. photoautotrophs
   e. all of the above

6. Figure 2.4 in your textbook suggests which kingdom had the first occurrence of multicellularity?
   a. Bacteria
   b. Archaea
   c. Protista
   d. Plantae
   e. Animalia

7. When groups of organisms share characteristics because they evolved the same features independently, rather than through common descent, we call these characteristics
   a. systematic.
   b. convergent.
   c. ancestral.
   d. derived.
   e. evolutionary.

8. Viruses can reproduce, evolve, and possess DNA. This makes viruses "alive." (True or False)

9. There are three main shapes of viral strains. They include
   a. bacillus, cocci, and helical.
   b. bacillus, icosahedral, and complex.
   c. cocci, rod, and complex.
   d. helical, icosahedral, and cocci.
   e. icosahedral, complex, and helical.

10. DNA technology was the primary tool used to determine that the Archaea belonged to a different evolutionary group from the Bacteria. (True or False)

11. Match each term with the best description.
    ___ horizontal gene transfer
    ___ phylum
    ___ evolutionary tree
    a. pictorial representation of relatedness
    b. a large group of closely related classes of organisms
    c. unexpected movement of DNA between separated lineages

12. In the Linnaean hierarchy, what are the units of classification, from smallest to largest?
    a. species, genus, order, family, phylum, class, kingdom
    b. kingdom, species, order, genus, family, class, phylum
    c. species, genus, family, order, class, phylum, kingdom
    d. phylum, family, genus, species, order, kingdom, class
    e. family, species, genus, order, kingdom, class, phylum

13. The name "*Homo sapiens*" reflects the species name and genus name for humans, in that order. (True or False)

Conceptual Understanding

1. A researcher has just discovered what she thinks is a new species. Initially, all she knows is that it is eukaryotic. To which of the kingdoms below could this new organism possibly belong?
   a. Animalia
   b. Protista
   c. Fungi
   d. Plantae
   e. any of the above

2. A researcher has just discovered what she thinks is a new species. Initially, all she knows is that it is eukaryotic. Further analysis reveals the organism in question is also multicellular and relies on other organisms for a source of carbon. Which kingdom(s) can now be excluded for this organism?
   a. Animalia
   b. Protista
   c. Fungi
   d. Plantae
   e. more than one of the above

3. The opposable thumb shared by humans, chimpanzees, and pandas is a product of convergent evolution among these species. (True or False)

4. Of all organisms, bacteria thrive in the broadest range of habitats. (True or False)

5. Which of the following would be considered a key medical consequence resulting from the discovery that fungi and animals are more closely related than previously thought?
   a. Antibiotics used to treat bacterial infections are ineffective against fungal infections.
   b. Yeast (a type of fungus) could hold the key to curing Lou Gehrig's disease.
   c. Fungal infections have traditionally been very difficult to cure because drugs that harm fungi typically have adverse effects on the human body.
   d. The research of fungi may someday allow scientists to discover new treatments for several genetic diseases.
   e. all of the above

6. Which of the following observations would be explained by the hypothesis of horizontal gene transfer?
   a. Bacterial DNA has been found in the cells of organisms in the Eukarya domain.
   b. Genes may be passed horizontally between siblings spontaneously.
   c. Shared derived features may evolve independently in separate lineages.

   d. Organisms may be classified in more than one grouping in the Linnaean hierarchy.
   e. all of the above

7. The discovery of a fossil dinosaur named *Oviraptor* at its nest has confirmed the view of many paleontologists that some dinosaurs exhibited parental care. This is an example of
   a. the power of convergent evolution.
   b. the importance of using shared, derived features to make evolutionary trees.
   c. using evolutionary trees to predict the biology of organisms.
   d. how DNA information is critical to generating evolutionary trees.
   e. poor science.

8. Viruses are essentially a particle of genetic information wrapped in a protein coat. Despite the fact that these particles are nonliving, they display certain characteristics of living organisms, including the ability to evolve. Which of the following is an example of the evolutionary capability of viruses?
   a. Viruses are capable of evading the body's natural defenses.
   b. Certain viruses use RNA as genetic material.
   c. Viruses are capable of reproducing rapidly throughout the body.
   d. Viruses are capable of infecting cells throughout the body.
   e. all of the above

9. Outside of the animal kingdom, our closest evolutionary relatives are probably in the kingdom Protista. (True or False)

10. Fungi are more closely related to plants than they are to animals. (True or False)

11. Flight is a characteristic shared among organisms from different groups. Both birds (class Aves) and bats (class Mammalia) are vertebrates that have the ability to use their front limbs to fly. Thus, flight is an example of
    a. a shared, derived feature.
    b. a Linnaean feature.
    c. convergent evolution.
    d. horizontal gene transfer.
    e. There is not enough information to answer the question.

12. If two organisms belong to the same class in the Linnaean classification hierarchy, it is also true that these organisms belong in
    a. the same genus and species.
    b. the same kingdom and phylum.
    c. the same family.

d. the same order.

e. a different kingdom and phylum.

RELATED ACTIVITIES

- There are no vaccines for viruses yet . . . or are there? Think of the vaccines you have received since you were born. Make a list of 10 vaccines, then determine what they were used against and if those illnesses could be caused by a virus. If viruses are not alive, then how could a vaccine be made for a virus? For a real challenge, determine how a vaccine can be made against an ever-evolving illness such as the flu. How does a vaccine work?

- Look ahead in our textbook to Section 20.7 on human evolution. Skim through the section and list as many primate or human characteristics you can find that would be useful for assembling a human evolutionary tree. Explain why the traits you have listed may have value in reconstructing human evolution.

- Look up a recent news event, either local or worldwide, that relates to a bacterial breakout (for example, mad cow disease). Find information on that particular bacterium, including its shape name, what other diseases or disorders it can cause, and how it affects the organism it infects. Then determine if this bacterium could have either a direct or indirect effect on the human race.

ANSWERS AND EXPLANATIONS

Factual Knowledge

1. d. For more information, see Section 2.1, *The extent of Earth's biodiversity is unknown.*

2. a. Bacteria are prokaryotes, which by definition means they lack internal membrane–bound organelles. All other choices are universal properties of life. For more information, see Section 2.1, *The Eukarya are sorted into four different kingdoms* (generally). See also, more specifically, Section 2.3, *Prokaryotes represent biological success.*

3. e. Bacteria and Archaea take advantage of all these characteristics to be successful. For more information, see Section 2.3, *Prokaryotes represent biological success* and *Prokaryotes occupy a great diversity of habitats.*

4. False. Recall that convergent features are similarities that exist between groups of organisms but do not necessarily indicate relatedness. DNA analysis provides a much more powerful and accurate tool for determining relatedness. For more information, see Section 2.1, *The Eukarya are sorted into four different kingdoms.*

5. e. Prokaryotes have the ability to obtain energy from other organisms and from the sun. They can also obtain their carbon from other organisms and from inorganic substances. Therefore, all of the terms apply to the prokaryotes. For more information, see Section 2.3, *Prokaryotes are unrivaled in metabolic diversity.*

6. c. Although some degree of multicellularity is found in all kingdoms except Bacteria and Archaea, the kingdom Protista is the most primitive and has the fewest multicellular species, suggesting that it is the ancestry of the other multicellular eukaryotes. For more information, see Figure 2.4 and Section 2.1, *The Eukarya are sorted into four different kingdoms.*

7. b. Convergent evolution is a process that can produce similar characteristics in organisms as a solution to environmental challenges, rather than through common descent. It can thus be a problem for systematists trying to determine evolutionary relationships on the basis of shared characteristics alone. For more information, see Section 2.1, *Evolutionary divergence explains the diversity of life on Earth.*

8. False. Refer to Chapter 1 for *all* of the characteristics of living organisms. Although viruses can evolve, reproduce, and possess DNA, they do not have all the characteristics of living organisms, and are essentially DNA in a protein coat. For more information, see Section 2.4, *Viruses lack cellular organization.*

9. e. The three main shapes of viruses include helical, icosahedral, and complex. The other shapes are classifications of bacteria. For more information, see Section 2.4, *Viruses are classified according to structure and type of infection,* and Figure 2.14.

10. False. Recall from the text that DNA studies initially alerted scientists to the existence of the Archaea domain. For more information, see Section 2.3, *Archaeans constitute a distinct domain of life.*

11. c. horizontal gene transfer. For more information, see Section 2.3, *Prokaryotes reproduce asexually.*
    b. phylum. For more information, see Section 2.2, The Linnaean System of Biological Classification.
    a. evolutionary tree. For more information, see Section 2.1, *A common origin explains the unity of life on Earth.*

12. c. Recall the catchphrase "King Phillip Cleaned Our Filthy Gym Shorts" to help you remember the correct order. For more information, see Section 2.2, The Linnaean System of Biological Classification.

13. False. In a scientific name, the first name is the genus (*Homo*) and the second is the species (*sapien*). For more information, see Section 2.2, The Linnaean System of Biological Classification.

Conceptual Understanding

1. e. All of the kingdoms listed contain eukaryotic organisms. For more information, see Section 2.1, *The Eukarya are sorted into four different kingdoms.*

2. d. Only plants can be excluded, because they are photosynthetic and produce their own food. The other three kingdoms all contain at least some multicellular (or colonial) species that must rely on other organisms for their carbon source. For more information, see Section 2.1, *The Eukarya are sorted into four different kingdoms.*

3. True. This is a classic example of convergent evolution in which the environmental benefits of an opposable thumb have forced common solutions to similar problems faced by the ancestors of these species. For more information, see Section 2.1, *Evolutionary divergence explains the diversity of life on Earth.*

4. True. Recall that habitat preferences in bacteria are diverse, including extremes of heat and cold, acidity, and salt concentration. For more information, see Section 2.3, *Archaeans constitute a distinct domain of life*, and Table 2.1, Lifestyles of the Extreme.

5. e. The realization that fungi are more closely related to animals than to plants has allowed scientists to discover new medically important links between these two groups. Refer to Figure 2.4 to see how these groups are located on the tree of life. For more information, see Section 2.1, *The Eukarya are sorted into four different kingdoms.*

6. a. The discovery that DNA from certain organisms can appear as a grab bag of DNA from across the tree of life has been explained by the horizontal gene transfer theory. For more information, see Section 2.3, *Prokaryotes reproduce asexually.*

7. c. Because birds and certain reptiles show parental care, and because both are related to dinosaurs by virtue of other shared, novel features, a prediction can be made that dinosaurs likely also showed parental care. For more information, see Section 2.1, *Evolutionary divergence explains the diversity of life on Earth.*

8. a. Viruses are capable of rapid reproduction once they have infected a host cell. During this process, viruses may undergo rapid evolution, which allows them to better evade the body's defense system. As an example, the influenza virus (the causative agent of the flu) changes or evolves each year. This is why revaccination is required on an annual basis. For more information, see Section 2.4, *Flu viruses evolve rapidly.*

9. False. A closer evolutionary relationship exists between the fungal and animal kingdoms. For more information, see Figure 2.4 and Section 2.1, *The Eukarya are sorted into four different kingdoms.*

10. False. Using DNA techniques, fungi have recently been found to be more closely related to animals than to plants. For more information, see Section 2.1, *The Eukarya are sorted into four different kingdoms.*

11. c. Flight is an example of convergent evolution in birds and bats, as this feature would have evolved independently in both groups. For more information, see Section 2.1, *Evolutionary divergence explains the diversity of life on Earth.*

12. b. If two organisms are in the same class, then the only thing that can be said for certain is that they are in the same kingdom and phylum. Refer to Figure 2.5 to see these relationships. For more information, see Section 2.2, The Linnaean System of Biological Classification.

CHAPTER 3 | Protista, Plantae, and Fungi

GETTING STARTED

Below are a few questions to consider before reading Chapter 3. These questions will help guide your exploration and assist you in identifying some of the key concepts presented in this chapter.

1. What can the process of river meandering teach us about interactions between living and nonliving components of an ecosystem?

2. What are the key features of the eukaryotes that distinguish them from the prokaryotes?

3. What is red tide, how does it affect humans, and which group of organisms is responsible for causing it?

4. What was the adaptation that enabled plants to grow to the towering heights seen among the trees?

5. Truffles, which can sell for more than $600 per pound, are a member of which Linnaean kingdom?

6. Why does the destruction of redwood forests contribute to water shortages in northern California?

7. What does humankind lose every time another species becomes extinct?

A GUIDE TO THE READING

The following concepts typically give students the most difficulty when exploring the content in Chapter 3 for the first time. For each concept, one or more references have been identified that may help you gain a better understanding of these potentially problematic areas.

Prokaryotes vs. Eukaryotes

As we learned in Chapter 2, organisms can be classified according to several schemes, one of which categorizes organisms according to cell structure. This scheme distinguishes cells on the basis of whether compartmentalized structures called "organelles" are present (eukaryote) or absent (prokaryote). Keep in mind that despite the similarity between the term "eukaryote" and the domain name "Eukarya," these refer to different levels of classification. The term "eukaryote" can be used to describe the structure of an organism's cells (organelles present), whereas the domain Eukarya refers to the group of organisms represented by four kingdoms (Protista, Plantae, Fungi, and Animalia).

For more information on this concept, be sure to focus on

- Section 3.1, *Eukaryotes have subcellular compartmentalization and larger cells*

The Protists—Evolution of the Eukaryote

The kingdom Protista is considered the most ancient group among the Eukarya domain. As discussed in the chapter, this group is extremely diverse in terms of size, shape, and lifestyle. As such, it is a group that has given scientists great difficulty in determining evolutionary relationships among its members. This struggle is illustrated by the evolutionary tree presented in Figure 3.6 of your textbook. Protists are placed in several groups and supergroups whose relationships are not fully understood. Some protists, such as red and green algae, are more closely related to plants than to other protists. One thing is certain, however: the protists were one of the first groups to develop multicellularity (the ability to live as members of a multicellular body) and sex (the contribution of DNA from both parents to offspring).

For more information on this concept, be sure to focus on

- Section 3.1, *Multicellularity evolved independently in several eukaryotic lineages*
- Section 3.1, *Sexual reproduction increases genetic diversity*
- Section 3.2, *Protists are not a natural grouping*
- Figure 3.6, The Protista

Angiosperms and Flowers

As discussed in the chapter, the development of flowers is a relatively recent evolutionary event within the plant kingdom. Flowering plants (angiosperms) produce seeds that are characteristically surrounded by a protective tissue that may take the form of a type of fruit. What defines an angiosperm, however, is the flower—a structure used for reproduction. What may be hard to understand is that a flower typically contains both male and female reproductive structures, as illustrated in Figure 3.13. Male reproductive structures form pollen containing sperm (male gametes), which can be carried by either wind or an animal to a nearby flower of the same species. The female reproductive structure, typically in the center of the flower, consists of the stigma (the site of pollen attachment), the style (carries pollen), and the ovary (contains the female eggs inside ovules). When pollen reaches an ovary, the result is a plant embryo that can take the form of a seed. Because both female and male reproductive structures are present in the same flower, an angiosperm is capable of self-pollination, or mating with itself.

For more information on this concept, be sure to focus on

- Section 3.3, *Angiosperms produce flowers and fruit*
- Figure 3.13, The Flower

The Fungi, a Group with Many Roles

As discussed in the chapter, fungi are described as playing many key roles in most ecosystems. The ability to play different roles stems from this group's ability to obtain nutrients and energy from a variety of sources and in a variety of ways. Fungi are capable of serving as a decomposer in an ecosystem by feeding on dead or dying organisms. Their role is to return vital nutrients back to the ecosystem for use by the system's producers. In a similar fashion, some groups of fungi may specialize in living on other organisms. This relationship can either be beneficial (mutualism) or detrimental (parasitism) to the host. Keep in mind that these relationships may or may not be based on the supply of nutrients. Therefore, there is no correlation between the terms "decomposer," "consumer," and "producer" and the terms "mutualist" and "parasite."

For more information on this concept, be sure to focus on

- Section 3.4, Fungi: A World of Decomposers
- Section 3.5, Lichens and Mycorrhizae: Collaborations between Kingdoms

Why Is Biodiversity Important?

Although the loss of a single spider or frog species may seem insignificant in the grand scheme of things, it is crucial to understand how important biodiversity is to the health of the ecosystems on our planet. The text describes research conducted by scientists that supports the notion that the greater the diversity of species present in an ecosystem, the heartier the ecosystem is, better able to make the most of available resources and recover after environmental stress. In addition, the biodiversity present in an ecosystem can help reduce invasion by foreign species and the potentially devastating competition that may result. Furthermore, humans depend on a wide variety of goods and services provided by the vast number of species in the biosphere. We are still discovering new and potentially beneficial products obtained from plant and animal species alike. Wouldn't it be a shame if the extinction of a particular plant species occurred before the discovery of its beneficial use as a new anticancer drug?

For more information on this concept, be sure to focus on

- Biology Matters, "The Importance of Biodiversity"

TYING IT ALL TOGETHER

Several concepts presented in this chapter will be revisited and discussed in greater detail in subsequent chapters, including

Eukaryotic Cell Structure

- Chapter 6—Section 6.4, Internal Compartments of Eukaryotic Cells

Photosynthesis

- Chapter 6—Section 6.4, *Chloroplasts capture energy from sunlight*
- Chapter 9—Section 9.3, Photosynthesis: Capturing Energy from Sunlight

Chloroplasts

- Chapter 6—Section 6.4, *Chloroplasts capture energy from sunlight*
- Chapter 9—Section 9.3, *Chloroplasts are photosynthetic organelles*

Sexual Reproduction

- Chapter 36—Section 36.3, Producing the Next Generation: Flower Form and Function

Plant Pollination

- Chapter 36—Section 36.3, Producing the Next Generation: Flower Form and Function

Biodiversity

- Chapter 19—Section 19.4, Speciation: Generating Biodiversity

PRACTICE QUESTIONS

Factual Knowledge

1. The group of organisms called "protists" was the first to utilize sexual reproduction. (True or False)

2. The evolutionary record suggests that the first occurrence of multicellularity was in the kingdom
   a. Archaea.
   b. Protista.
   c. Fungi.
   d. Plantae.
   e. Animalia.

3. The common characteristic of red tides, malaria, and potato blight is that they are all
   a. caused by bacteria.
   b. caused by protists.
   c. caused by fungi.
   d. diseases of plants.
   e. none of the above

4. All protists, like bacteria, are microscopic, single-celled organisms. (True or False)

5. Which of the following statements applies to all protists?
   a. They are autotrophic.
   b. They are pathogens.
   c. They are multicellular.
   d. They contain subcellular compartments.
   e. They only reproduce sexually.

6. Extensive root systems and the presence of a waxy cuticle are among the reasons plants have become so successful on land. (True or False)

7. Consider the photographs of various plants in Figure 3.15. Which of the following statements about these organisms is false?
   a. They are photoheterotrophs.
   b. They are multicellular photosynthesizers.
   c. Several produce flowers and seeds.
   d. They function as producers in food webs.
   e. Each has a waxy cuticle.

8. Which one of the following is *not* found in both the gymnosperm and angiosperm plants?
   a. well-developed vascular systems for internal transport
   b. extensive root systems for acquiring water and nutrients
   c. seeds that disperse from the parent and thus increase offspring survival
   d. sexual reproduction using special structures called "flowers"
   e. products that are useful for human consumption

9. Fungi are important to food webs because they
   a. photosynthesize.
   b. have prokaryotic cell structure.
   c. serve as key decomposers of dead organic matter.
   d. generate antibiotic chemicals that kill competing plants and animals.
   e. form mutually beneficial relationships with other organisms.

10. Consider the fungi shown in Figures 3.16 through 3.20. What do all of these organisms have in common?
    a. a mycelium
    b. many individual hyphae
    c. chemoheterotrophy
    d. spore production
    e. all of the above

11. Maintaining biodiversity is a primary key to a healthy ecosystem. When ecosystems are healthy, humans can harvest a wide variety of useful materials. Which of the following useful products comes from nonhuman species?
    a. one-fourth of all prescription drugs
    b. the oxygen we breathe
    c. meat and dairy products
    d. vegetables and wood
    e. all of the above

12. A lichen is best described as
    a. a sexually reproducing plant.
    b. a mutualistic relationship.
    c. a primitive prokaryote.
    d. a pathogenic fungus.
    e. an invasive species.

13. Association with mycorrhizae can be highly beneficial to a plant's health, growth, and productivity because mycorrhizae
    a. are spongy and hold lots of water in the soil.
    b. make use of extra sugar that would otherwise harm the plant.
    c. help a plant obtain additional nutrients from the soil.

d. prevent any other plants from growing nearby.

e. perform photosynthesis to produce sugars.

Conceptual Understanding

1. A researcher has just discovered what she thinks is a new species. Initially, all she knows is that it is eukaryotic. To which of the kingdoms below could this new organism possibly belong?
   a. Animalia
   b. Protista
   c. Fungi
   d. Plantae
   e. any of the above

2. Further analysis reveals the organism in question 1 is also multicellular and relies on other organisms for a source of carbon. Which kingdom or kingdoms can now be excluded for this organism?
   a. Animalia
   b. Protista
   c. Fungi
   d. Plantae
   e. more than one of the above

3. A population of fungi that reproduce sexually is more likely than an asexual population to evolve following resistance exposure to an antifungal chemical. (True or False)

4. The protists make up a relatively small kingdom of prokaryotic organisms whose evolutionary relationships to the fungi, plants, and animals are well understood. (True or False)

5. Both plantlike protists and plants
   a. are prokaryotic.
   b. have some cellular organelles in common.
   c. serve as decomposers in an ecosystem.
   d. evolved after the first animals appeared.
   e. were among the last multicellular groups to evolve.

6. Vascular systems were a key innovation in plant evolution because they
   a. allowed plants to become taller and more efficient at nutrient transport.
   b. made possible the production of oxygen during photosynthesis.
   c. increased the success rate of sexual reproduction.
   d. were needed before photosynthesis could evolve.
   e. all of the above

7. Which of the following plants would be least able to survive in a dry environment?
   a. sword fern
   b. cycad
   c. spruce tree
   d. tomato plant
   e. rose bush

8. The success of the angiosperms is due largely to an evolutionary innovation—the flower, which dramatically increases the efficiency of sexual reproduction. (True or False)

9. Select all of the statements below that correctly apply to the kingdom Fungi.
   a. The fungi are all eukaryotic and multicellular and rely on other organisms for food.
   b. Some fungi form beneficial interrelationships with plants.
   c. The fungal life cycle typically includes a spore stage.
   d. Certain fungi are natural sources of antibiotic substances.
   e. Fungi can form close associations with protists.

10. Which of the following natural processes would probably be most severely affected if all fungi suddenly disappeared?
    a. decomposition
    b. photosynthesis
    c. bacterial reproduction
    d. viral reproduction
    e. evolutionary innovation

11. What do Pacific yew trees, coast redwoods, reeds, and pineapple all have to do with the issue of whether humans should work to maintain biodiversity?
    a. Each is regarded as a key species in its environment.
    b. They are all endangered species.
    c. They all provide important products and services that are used by people.
    d. These species are known to have survived previous mass extinctions.
    e. They are all the source of important medicines.

12. A farmer notices a significant decrease in crop growth after application of a fungicide to kill a fungal pathogen in the soil. The most likely explanation is that the fungicide
    a. was also harmful to the plants because they are very closely related to fungi.
    b. killed fungi that were providing the plants with sugar.
    c. killed lichens that were performing photosynthesis to provide sugars for the plants.
    d. was not strong enough to kill the pathogen.
    e. killed mycorrhizal fungi that were helping the plants get water and minerals.

RELATED ACTIVITIES

- Choose two examples of organisms from different kingdoms discussed in this chapter. List as many key biological features the two have in common as you can think of. (Hint: Think back to the ideas discussed in Chapter 1.) Then expand your list to include the characteristics that are different about these organisms.
- Imagine that an elementary school child comes to you with a question about biology. As part of an assignment about the different kinds of organisms, he must classify as many types of living creatures as he can. Your friend is having trouble figuring out what to do with the lichens and viruses. Compose a simple explanation of how these two types of organisms are viewed in the Linnaean classification hierarchy. (Refer to Chapter 2.)
- Find three different sites on the Internet that address the issue of biodiversity. Write a one-page essay that summarizes how the concept of biodiversity is relevant to the organization maintaining the web page, and compare the approach each takes to support (or refute) the need for biodiversity.

ANSWERS AND EXPLANATIONS

Factual Knowledge

1. True. The exchange of DNA with another organism to pass it to offspring was an adaptation that first appeared in the protists. For more information, see Section 3.1, *Sexual reproduction increases genetic diversity.*
2. b. Although some degree of multicellularity is found in all kingdoms except Bacteria and Archaea, the kingdom Protista is the most primitive and has the fewest multicellular species, suggesting that it is the ancestor of the other multicellular eukaryotes. For more information, see Section 3.1, *Multicellularity evolved independently in several eukaryotic lineages.*
3. b. All three of these conditions are caused by protists. Some protists are pathogens. For more information, see Section 3.2, Protista: The First Eukaryotes.
4. False. Although the large majority of protists are single-celled and microscopic, notable exceptions—such as slime molds and seaweeds—exist. For more information, see Section 3.2, *Most protists are single-celled and microscopic.*
5. d. Protists are all eukaryotes that have membrane-enclosed compartments in their cells. The other traits apply to some, but not all, protists. For more information, see Section 3.1, *Eukaryotes have subcellular compartmentalization and larger cells.*

6. True. The root system provides water and nutrients, and the waxy cuticle helps prevent desiccation. For more information, see Section 3.3, *Plants had to adapt to life on land* and *The vascular system enables plants to move fluids efficiently.*
7. a. Although plants are capable of deriving their energy from sunlight, they are considered photoautotrophs because they extract carbon from carbon dioxide in the air. In contrast, photoheterotrophs derive their carbon from other organisms or inorganic objects rather than carbon dioxide. For more information, see Section 3.3, Plantae: The Green Mantle of Our World.
8. d. Flowers are found in the angiosperms only. Instead of flowers, gymnosperms form cones or conelike structures for protection of their seeds. For more information, see Section 3.3, *The evolution of seeds contributed to the success of gymnosperms* and *Angiosperms produce flowers and fruit.*
9. c. The first two choices are not correct for fungi, which are eukaryotes that cannot carry out photosynthesis. The last two choices, although true of fungi, are not what make fungi particularly important in food webs. The main role of fungi is decomposition. For more information, see Section 3.4, *Fungi play a key role as decomposers.*
10. e. All of these are general features of fungal biology. For more information, see Section 3.4, *Fungi are adapted for absorptive heterotrophy* and *Fungi have unique ways of reproducing.*
11. e. All these are products that come from nonhuman species. Therefore, the more diverse an ecosystem, the more usable products may be derived from that system. For more information, see the Biology Matters box, "The Importance of Biodiversity."
12. b. Lichens are created by a mutualistic relationship between a fungus and a photosynthetic microorganism in which both partners benefit from the interaction. For more information, see Section 3.5, *Lichens contain a fungus and a photosynthetic microbe.*
13. c. Fungal mycelia are very thin and highly branched, providing much surface area to absorb water and nutrients that are shared with the plant root. For more information, see Section 3.5, *Mycorrhizae are beneficial associations between a fungus and plant roots.*

Conceptual Understanding

1. e. All the kingdoms listed contain eukaryotic organisms. For more information, see the Introduction in Chapter 3, The Mind-Boggling Diversity of Life.
2. d. Only the plants can be excluded, as they are photosynthetic and produce their own food. The other three kingdoms all contain at least some multicellular (or

colonial) species that must rely on other organisms for their carbon source. For more information, see Section 3.2, *Protists are autotrophs, heterotrophs, or mixotrophs*; Section 3.3, Plantae: The Green Mantle of Our World; and Section 3.4, Fungi: A World of Decomposers.

3. True. Sexual reproduction is a major source of genetic diversity that allows a population to evolve in response to stress, whereas asexual reproduction makes identical clones. For more information, see Section 3.1, *Sexual reproduction increases genetic diversity*.

4. False. Recall that protists are eukaryotic and that the evolutionary relationship between the protists and the other eukaryotic kingdoms is poorly understood. For more information, see Section 3.2, *Protists are not a natural grouping*.

5. b. Plants and plantlike protists are both eukaryotic, so similar cellular organelles are to be expected. The decomposers of an ecosystem typically are fungi. Recall that protists represent an early stage in eukaryotic and multicellular evolution, so the last two answers are also incorrect. For more information, see Section 3.1, *Eukaryotes have subcellular compartmentalization and larger cells*.

6. a. Plant vascular tissue makes long-range transport of nutrients and water possible. The plant body can thus become larger. The other choices are incorrect because the more primitive nonvascular plants also photosynthesize and reproduce sexually with great success. For more information, see Section 3.3, *The vascular system enables plants to move fluids efficiently*.

7. a. All of the other plants on the list are either gymnosperms or angiosperms whose sperm is dispersed in pollen, whereas ferns have flagellated sperm that must swim through water to reach an egg. For more information, see Section 3.3, *The evolution of seeds contributed to the success of gymnosperms*.

8. True. The evolutionary innovation of using flowers for reproduction was so successful that it caused an explosion of new species during plant evolution. For more information, see Section 3.3, Angiosperms produce flowers and fruit.

9. b, c, d, e. Recall that some fungi are unicellular (for example, yeast), and all of them absorb food produced by other organisms. The close associations formed with plants (mycorrhizae) and protists (lichens) prove beneficial for both groups. For more information, see Section 3.4, *Fungi are adapted for absorptive heterotrophy*; *Fungi have unique ways of reproducing*; *Fungi can benefit human society*; and Section 3.5, Lichens and Mycorrhizae: Collaborations between Kingdoms.

10. a. Decomposition of organic material and the return of key nutrients to the nonliving portion of the biosphere are the two main roles that fungi play. For more information, see Section 3.4, *Fungi play a key role as decomposers*.

11. c. The best answer is that all four are examples of species that produce valuable products or services for human use. Some provide medicine or food, whereas others have important roles in mediating the impacts of weather on their local environments. For more information, see the Biology Matters box, "The Importance of Biodiversity."

12. e. Mycorrizhae are often killed by commercial agriculture practices, and this decreases the efficiency with which crop plants obtain water and minerals from the soil. For more information, see Section 3.5, *Mycorrhizae are beneficial associations between a fungus and plant roots*.

CHAPTER 4 | Animalia

GETTING STARTED

Below are a few questions to consider before reading Chapter 4. These questions will help guide your exploration and assist you in identifying some of the key concepts presented in this chapter.

1. What can we learn about human biology from microscopic organisms like choanoflagellates?

2. What features define an organism as an animal?

3. What traits distinguish animals from the other kingdoms of eukaryotes?

4. What adaptations allow reptiles to be successful in dry climates?

5. How are birds specialized for flight beyond just having wings?

6. Which group of animals dominates Earth in terms of number of species?

A GUIDE TO THE READING

The following concepts typically give students the most difficulty when exploring the content in Chapter 4 for the first time. For each concept, one or more references have been identified that may help you gain a better understanding of these potentially problematic areas.

Animal Body Cavities

One of the evolutionary innovations that helped animals develop the extreme diversity displayed by this group was the formation of a complete body cavity. The body cavity is defined as an interior space with openings at either end (mouth, anus). The text describes two distinct evolutionary lineages with a complete body cavity: the protostomes and deuterostomes. These two groups are distinguished by which of the two openings develops into the adult's mouth. In the protostomes (insects, worms, snails), during development the opening that will become the mouth forms first. In the deuterostomes (echinoderms and vertebrates, including humans), the opening that becomes the adult anus forms first during the organism's development. This distinction is important to understand as we learn more about organismal growth and development in subsequent chapters.

For more information on this concept, be sure to focus on

- Section 4.2, *Animals exhibit unique patterns of embryo development*
- Section 4.2, *Some animals evolved complex body cavities*

Mass Extinction

Many scientists believe that we are currently undergoing a mass extinction event driven by the presence of humans and an explosion in the human population. Although it is difficult to determine the exact extinction rate occurring around the globe, scientists have documented that more than 20 percent of the freshwater fish and bird species that existed 2,000 years ago are now extinct. The primary difference between this mass extinction event and those that occurred previously is the influence of humankind.

For more information on this concept, be sure to focus on

- Biology Matters, "Goodbye, Catch of the Day?"

TYING IT ALL TOGETHER

Several concepts presented in this chapter will be revisited and discussed in greater detail in subsequent chapters, including

Sexual Reproduction and Animal Development

- Chapter 33—Section 33.1, Sexual and Asexual Reproduction in Animals
- Chapter 33—Section 33.3, Human Reproduction: From Fertilization to Birth
- Chapter 33—Section 33.6, How Development Is Controlled

Animal Behavior

- Chapter 34—Animal Behavior

Animal Evolution

- Chapter 19—Section 19.1, Adaptation: Adjusting to Environmental Challenges
- Chapter 20—Section 20.1, The Fossil Record: A Guide to the Past
- Chapter 20—Section 20.2, The History of Life on Earth
- Chapter 20—Section 20.6, The Origin and Adaptive Radiation of Mammals

Mass Extinction

- Chapter 20—Section 20.4, Mass Extinction: Worldwide Losses of Species

PRACTICE QUESTIONS

Factual Knowledge

1. Of the animal groups listed below, which one evolved first?
   a. flatworms
   b. sponges
   c. arthropods
   d. vertebrates
   e. mollusks

2. Because of their ingestive mode of nutrition, animals function as _____ in ecosystems, but never as _____.
   a. consumers and decomposers; producers
   b. producers; consumers and decomposers
   c. producers and consumers; decomposers
   d. decomposers; producers and consumers
   e. decomposers and producers; consumers

3. Organ systems are found only in recently evolved deuterostomes, including chordates and vertebrates. (True or False)

4. The advantage of a segmented body plan is that it allows for
   a. more internal space.
   b. redundancy.
   c. increased surface area.
   d. specialized functions.
   e. longer body length.

5. Tissues and symmetry are found in all animals except sponges. (True or False)

6. Cnidarians are a group of animals that have
   a. neither tissues nor organs.
   b. organs but no tissues.
   c. two tissues but no organs.
   d. three tissues but no organs.
   e. both tissues and organs.

7. Which of the following traits describe(s) protostomes?
   a. bilateral symmetry at some point in the life cycle
   b. three embryonic cell layers
   c. embryonic mouth developing from the blastopore
   d. ventral nervous system with an anterior brain
   e. all of the above

8. Earthworms have complex organ systems for digestion and circulation. (True or False)

9. Which phylum of animals contains the greatest number of species among eukaryotes?
   a. annelids
   b. mollusks
   c. arthropods
   d. echinoderms
   e. chordates

10. Which of the following traits describe(s) deuterostomes?
    a. bilateral symmetry at all life stages
    b. dorsal notochord
    c. embryonic mouth develops from the blastopore
    d. dorsal hollow nerve cord
    e. all of the above

11. Which of the following is *not* a function of the echinoderm's water vascular system?
    a. regeneration
    b. gas exchange
    c. movement
    d. feeding
    e. providing suction for tube feet

12. Which of the following are traits shared by all vertebrates? (Select all that apply.)
    a. jaws
    b. closed circulatory system
    c. vertebral column
    d. bony skeleton
    e. anterior braincase

13. Adaptations that provide an advantage in drier terrestrial habitats are seen in
    a. amphibians but not reptiles.
    b. reptiles but not amphibians.
    c. birds but not reptiles.
    d. mammals but not birds.
    e. all tetrapods.

14. Which of the following animals shows an embryonic developmental pattern that classifies it as a deuterostome?
    a. bird
    b. snail
    c. insect
    d. worm
    e. none of the above

Conceptual Understanding

1. Evolutionarily speaking, to which of the following animal groups are you most closely related?
    a. starfish
    b. flatworms
    c. snails
    d. birds
    e. insects

2. Which of the following animals is most closely related to an octopus?
    a. jellyfish
    b. sea anemone
    c. snail
    d. lobster
    e. snake

3. Jellyfish and people are both animals, and so they share a number of key biological features characteristic of the entire kingdom. However, because of their distant ancestry, these two groups also exhibit important differences in body structure and function. Which of the following characteristics is present in humans but *not* in jellyfish?
    a. true tissues
    b. multicellular organization
    c. organ systems
    d. ingestion of food
    e. mobility

4. Among the more advanced animals, there are two distinct ways in which the basic body tissues become arranged during the development of a fertilized egg. In _____, such as insects and earthworms, the first opening to form in the embryo becomes the mouth, whereas in _____, such as humans and sea stars, the mouth forms from the second embryonic opening.
    a. protostomes; deuterostomes
    b. deuterostomes; protostomes
    c. heterostomes; deuterostomes
    d. deuterostomes; heterostomes
    e. protostomes; heterostomes

5. Outside of the animal kingdom, our closest evolutionary relatives are probably in the kingdom Protista. (True or False)

6. Which of the following lifestyles would be most likely for an animal with radial symmetry?
    a. flight
    b. burrowing
    c. climbing trees
    d. sessile marine
    e. active swimming

7. Which of these is an example of a coelomate animal?
    a. sponge
    b. sea anemone
    c. roundworm
    d. flatworm
    e. sea cucumber

8. Consider an animal that has very different needs at different stages of its life—for example, one that is aquatic and carnivorous when young, then develops into an adult that is carnivorous and terrestrial. This animal can best be successful at both stages of its life by
    a. being coelomate.
    b. being a protostome.
    c. undergoing metamorphosis.
    d. molting.
    e. being ectothermic.

9. Imagine that you have discovered a new species resembling a tunicate. If you wanted to confirm that this animal is a chordate, which of the following traits would you look for?
    a. cephalization
    b. closed circulatory system
    c. embryonic mesoderm
    d. larval notochord
    e. adult coelom

10. Jawless fishes have been completely supplanted and replaced by fishes with a more advantageous hinged jaw. (True or False)

11. What is the adaptive advantage provided by both furs and feathers?
    a. protecting skin from sunlight
    b. waterproofing
    c. deterring predators
    d. regulating temperature
    e. providing sensory perception extended from the body

12. Which of the following is *not* an innovation specific to mammalian parental care?
    a. internal fertilization
    b. plancental embryo growth
    c. mammary glands
    d. marsupial pouch
    e. All of the above are mammalian innovations.

13. A recent revision of the evolutionary tree suggests that the _____ are the closest living relatives of the now extinct dinosaurs.
    a. amphibians
    b. birds
    c. mammals
    d. fungi
    e. reptiles

RELATED ACTIVITIES

• *Tiktaalik* and *Archaeopteryx* are two fossil discoveries that have been hailed as "missing links" in vertebrate evolution. Using library and Internet resources, learn more about one of these famous fossils. Where was it found and by whom? What important insights did this discovery provide into how vertebrates evolved? Compose a one-page essay on the history and significance of your chosen fossil.

• This chapter's Biology Matters box "Goodbye, Catch of the Day?" highlights concerns about human-driven extinction. Select a species at risk of extinction that you think is interesting, important, or both. Learn more about the species. Where does it live? What pressures are causing its decline? How are people contributing to those pressures? What effect would its extinction have on its ecosystem? On people? Prepare a poster to educate the public about this species and the dangers of extinction.

ANSWERS AND EXPLANATIONS

Factual Knowledge

1. b. Sponges are the most primitive animal group, being a relatively loose association of specialized cell types and lacking even tissue-level organization. For more information, see Section 4.1, The Evolutionary Origins of Animalia.

2. a. Producers capture energy directly from inorganic sources (for example, the sun) and convert it to organic energy, as the plants in a food web do during photosynthesis. The animals in a food web gather premade organic energy from other organisms, in the role of either consumers or decomposers. For more information, see Section 4.1, The Evolutionary Origins of Animalia.

3. False. Organ systems are found in most protostomes as well as all deuterostomes. For more information, see Section 4.2, *Most animals have organs and organ systems.*

4. d. Segmentation allows for repeated structures that can diverge over evolutionary time to gain specialized functions. For more information, see Section 4.2, *Segmentation enabled division of labor among body parts.*

5. True. Sponges have specialized cells, but they are not organized into true tissues. Sponges also lack organized body plans. All other animals have both true tissues and some form of symmetry. For more information, see Section 4.2, *Most animals have true tissues* and *Most animals have symmetrical bodies.*

6. c. Cnidarians, such as jellyfish and coral, have two distinct tissue layers but lack coherent organs. For more information, see Section 4.3, *Cnidarians and ctenophores display radial symmetry.*

7. e. All of these are characteristic traits shared by protostomes, although the group gets its name from the fate of the blastopore. For more information, see Section 4.4, The Protostomes.

8. True. Despite their external simplicity, earthworms have well-developed circulatory and digestive systems. For more information, see Section 4.4, *Annelids are coelemic worms with segmented bodies.*

9. c. Arthropoda—a diverse phylum including crustaceans, spiders, and insects—is the largest eukaryotic phylum, mostly because of the stunning success of insects. For more information, see Section 4.4, *Arthropoda is the most species-rich phylum.*

10. d. All deuterostomes have a dorsal hollow nerve cord, whereas notochords are found only in chordates, not in echinoderms. For more information, see Section 4.5, The Deuterostomes—I: Echinoderms, Chordates, and Relatives.

11. a. Although many echinoderms, such as sea stars, do have a remarkable ability to regenerate lost body parts, this is not a result of their water vascular system. Everything else on the list is. For more information, see Section 4.5, *Echinoderms use a water vascular system for locomotion and gas exchange.*

12. b, c, e. All vertebrates have a closed circulatory system with a heart and an internal skeleton that includes

a braincase and vertebral column. Jaws and bony skeletons were later evolutionary adaptations that developed among some, but not all, vertebrates. For more information, see Section 4.6, The Deuterostomes—II: Vertebrates and *Jaws and a bony skeleton represent major steps in vertebrate evolution.*

13. b. Reptiles developed several key adaptations, such as amniotic eggs and water-retaining kidneys, that make them successful in dry environments inhospitable to amphibious lifestyles. For more information, see Section 4.6, *Reptiles exhibit adaptations to a drier environment.*

14. a. Recall that in deuterostomes, the second opening that forms at either end of the body cavity eventually develops into the mouth. This group includes all vertebrates, such as the bird. For more information, see Section 4.6, The Deuterostomes—II: Vertebrates.

Conceptual Understanding

1. d. Humans are vertebrates, so we can eliminate the starfish, flatworms, snails, and insects from the list. The only other vertebrates on the list are the birds. For more information, see Section 4.6, The Deuterostomes—II: Vertebrates.

2. c. Snails (gastropods) and octopi (cephalopods) are both members of the diverse phylum of mollusks. For more information, see Section 4.4, *Mollusks constitute the largest marine phylum.*

3. c. Recall that the evolutionary development of organs and organ systems followed the development of tissues on the evolutionary tree. Cnidarians (jellyfish) do not contain organs or organ systems. All the other choices are found in both jellyfish and humans. For more information, see Section 4.3, *Cnidarians and ctenophores display radial symmetry.*

4. a. Recall that the terms "protostome" and "deuterostome" refer to the order of formation of the two openings on either side of the body cavity in animals. In protostomes, the first opening develops into the mouth, whereas the second opening forms the mouth in deuterostomes. For more information, see Section 4.2, *Animals exhibit unique patterns of embryo development.*

5. True. Recent DNA evidence suggests a close evolutionary relationship exists between the animal kingdoms and a type of protist called "choanoflagellates."

For more information, see the Applying What We Learned article, "Clues to the Evolution of Multicellularity."

6. d. Radial symmetry is most beneficial for an animal that is stationary because it allows uniform access to the environment in any direction. For more information, see Section 4.2, *Most animals have symmetrical bodies.*

7. e. Sea cucumbers are echinoderms, a group of deuterostomes. All deuterostomes have true coeloms, as do many protostomes but none of the other examples on this list. For more information, see Section 4.2, *Some animals evolved complex body cavities.*

8. c. Metamorphosis allows an animal to switch between very different physical forms. That would allow the animal described here (a toad) to be specially adapted for two different environments and lifestyles at different stages of its life. For more information, see Section 4.4, *Insects.*

9. d. Although all of these traits are present in chordates, only the dorsal notochord is specific to this group. Tunicates (sea squirts) lack a notochord as adults, but it is present in their larvae. For more information, see Section 4.5, *Chordates possess a dorsal notochord at some stage of their life cycle.*

10. False. Although hinged jaws are an adaptation that represents a significant evolutionary advance, allowing more efficient predation, jawless fishes like hagfish and lampreys are still alive and well. For more information, see Section 4.6, *Jaws and a bony skeleton represent major steps in vertebrate evolution.*

11. d. Furs and feathers both serve as insulation that retains heat for endothermic animals. For more information, see Section 4.6, *Birds are adapted for flight* and *Mammals can live in diverse habitats because they regulate body temperature.*

12. a. Although mammals do use internal fertilization, they are by no means the only group to do so; reptiles and birds also rely on internal fertilization for reproduction. For more information, see Section 4.6, *Parental care contributed to the success of mammals.*

13. b. Compelling evidence now exists to suggest that birds are direct descendants of the dinosaurs. Refer to the figure of the *Archaeopteryx* fossil on p. 101 for additional information on this discovery. For more information, see Section 4.6, *Birds are adapted for flight.*

CHAPTER 5 | The Chemistry of Life

GETTING STARTED

Below are a few questions to consider before reading Chapter 5. These questions will help guide your exploration and assist you in identifying some of the key concepts presented in this chapter.

1. What distinguishes an element's atomic number from its atomic mass number?

2. What type of chemical bond results from the sharing of electrons between two atoms?

3. What are the differences between a solution, a solute, and a solvent?

4. Why are nonpolar molecules considered to be hydrophobic?

5. What biological role do buffers play in regulating the pH of a solution?

6. What function does the carbohydrate glycogen play inside the animal cell?

7. What is the relationship between the primary, secondary, tertiary, and quaternary structures of a protein and its function?

8. Consumption of too many saturated fats can result in a variety of diseases. What is the primary difference between saturated and unsaturated fats?

A GUIDE TO THE READING

The following concepts typically give students the most difficulty when exploring the content in Chapter 5 for the first time. For each concept, one or more references have been identified that may help you gain a better understanding of these potentially problematic areas.

Radioisotopes

Each element is characterized by its atomic number, which is the number of protons found in the atom's nucleus. Recall that the nucleus also contains a certain number of neutrons as well. The number of neutrons usually equals the number of protons for a particular element, although the number of neutrons may vary. An isotope is an element with the same number of protons, but a different number of neutrons. The example described in the chapter is carbon, which can exist as C-12 (six protons plus six neutrons) or C-14 (six protons plus eight neutrons). Isotopes such as C-14 may be unstable as a result of having more than the normal number of neutrons. Unstable isotopes tend to give off energy (in the form of radiation) through the process of radioactive decay. As a result, these unstable isotopes are often referred to as "radioisotopes." High-energy radioisotopes can cause damage to DNA and result in cancer. Lower-energy radioisotopes can be useful for medical diagnostics because they can be used to trace molecules as they proceed through a biological process, serving as a chemical "tracer."

For more information on this concept, be sure to focus on

- Section 5.1, *Some elements exist in variant forms called isotopes*

Polar vs. Nonpolar Molecules

As we saw in the chapter, molecules can be held together through the sharing of electrons. When electrons are not shared equally between atoms, what results is the uneven distribution of electrical charge in the resulting molecule: half the molecule will have a slight positive charge, and the other half will carry a slight negative charge. Such molecules containing an uneven distribution of electrical charge are called "polar." A good example of a polar molecule is water. Polar molecules tend to be attracted to one another through the partial electrical charges that exist. The electrical attractions that keep these molecules close together are called "hydrogen bonds." Polar molecules, as a class, will be soluble in water because they can interact with the partial electrical charges of the water molecules. Molecules that have an affinity for interacting with water like this are called "hydrophilic," or "water-loving." In contrast, molecules that are held together by the equal sharing of electrons tend to have no partial electrical charge and are called "nonpolar." With no charges available to interact with partially charged water molecules, nonpolar molecules tend to be insoluble in water. A good example of a nonpolar molecule is a fat molecule, or lipid. These molecules are called "hydrophobic," or "water-fearing."

For more information on this concept, be sure to focus on

- Section 5.3, *Water is a polar molecule*
- Section 5.3, *Water is a solvent for charged or polar substances*
- Figure 5.8, Charged Substances Dissolve in Water to Form Solutions

pH Scale

The acidity of a solution is a measure of the solution's concentration of free hydrogen ions (H+). The higher the concentration of free hydrogen ions in a solution, the more acidic that solution will be. The lower the concentration of free hydrogen ions, the less acidic (and more basic) the solution will be. The pH scale was developed to describe how acidic a solution is by measuring the concentration of free hydrogen ions in a solution. The scale ranges from 0 to 14, with a lower pH value indicating a higher concentration of hydrogen ions in solution and a more acidic solution. At the other end of the scale, higher pH values indicate a lower concentration of hydrogen ions and a more basic solution. Most biological systems maintain a pH of around 7, which is considered neutral. Maintaining proper pH is vital to the survival of biological systems. To assist in this process, special compounds called "buffers" are used to absorb excess hydrogen ions when a solution becomes too acidic or to donate hydrogen ions when a solution becomes too basic. The key to understanding the relationship of pH and acidity is realizing that the pH scale is actually inverted in terms of hydrogen ion concentration (that is, low pH = high acidity; high pH = low acidity).

For more information on this concept, be sure to focus on

- Section 5.5, The pH Scale
- Figure 5.11, The pH Scale Indicates Hydrogen Ion Concentration, a Measure of Acidity versus Basicity

Saturated vs. Unsaturated Fats

Fatty acids are molecules that typically consist of long chains of carbon and hydrogen atoms. These long chains can take one of two forms. When each carbon atom in a chain is attached to carbon atoms on either side and two hydrogen atoms, the chain is said to be "saturated" or "full" of hydrogen atoms. Saturated fatty acid chains tend to be straight, which enables them to pack together tightly. Fat molecules consisting of mostly saturated fatty acids tend to be solid at room temperature. In contrast, fatty acid chains in which two or more of the carbon atoms are held together with double covalent bonds will have fewer than the maximum number of two hydrogen atoms per carbon. As a result, unsaturated fatty acid chains will have small bends, or "kinks," in their chain (refer to Figure 5.18 in your textbook). These kinks tend to prohibit the molecules from packing together tightly. Fats consisting of mostly unsaturated fatty acids will tend to be liquid at room temperature. The key to understanding the difference between saturated and unsaturated fatty acids is understanding that saturated fats have carbon chains that are "full" or "saturated" with hydrogen atoms, whereas the carbon chains of unsaturated fats will have fewer than the maximum number of hydrogen atoms bound to the chain.

For more information on this concept, be sure to focus on

- Section 5.9, Lipids
- Figure 5.18, Saturated and Unsaturated Fatty Acids Are the Two Main Types of Fatty Acids

TYING IT ALL TOGETHER

Several concepts presented in this chapter may be revisited and discussed in greater detail in subsequent chapters, including

Biological Membranes

- Chapter 6—Section 6.2, The Plasma Membrane

Energy

- Chapter 8—Section 8.1, The Role of Energy in Living Systems

Macromolecules and Nutrition

- Chapter 8—Section 8.2, Metabolism
- Chapter 27—Section 27.1, Nutrients That Animals Need

Chemical Reactions

- Chapter 8—Section 8.2, *Chemical reactions are governed by the laws of thermodynamics*

- Chapter 8—Section 8.3, *Enzymes increase reaction rates by lowering the activation energy barrier*

Radiation in Science

- Chapter 14—Section 14.4, *Normal gene function depends on DNA repair*

DNA

- Chapter 14—Section 14.2, The Three-Dimensional Structure of DNA

Building Biological Molecules

- Chapter 14—Section 14.3, How DNA Is Replicated
- Chapter 15—Section 15.5, Translation: Information Flow from mRNA to Protein

PRACTICE QUESTIONS

Factual Knowledge

1. The elements hydrogen, nitrogen, and carbon
   a. can all form ionic bonds with other elements.
   b. contain neutrons in their atomic nuclei.
   c. have numerous electrons that orbit the inner core of their atoms.
   d. are common elements in the molecules that make up living organisms.
   e. all of the above

2. If two electrons come close to one another, they
   a. are attracted by their opposite electric charges.
   b. are repelled by their like electric charges.
   c. form an element.
   d. form a hydrogen bond.
   e. produce an ion.

3. Which of the following statements about covalently bonded molecules is *false*?
   a. Electrons are shared between the atoms that make up the molecule.
   b. Noncovalent bonds may also be present, especially if the molecule is large.
   c. Opposing electrical charges hold the molecule together.
   d. The chemical bonds that hold the molecules together are relatively strong.
   e. Protons and electrons are present in the atoms that make up the molecule.

4. H_2O and CH_4 are both examples of
   a. molecules.
   b. ions.

   c. elements.
   d. acids.
   e. all of the above

5. Which of the following statements about water is true?
   a. It contains two of the most common elements on Earth.
   b. It is critical for many of the chemical processes found in living systems.
   c. Polar substances that can form hydrogen bonds will dissolve in it.
   d. Nonpolar substances like oils are not soluble in it.
   e. all of the above

6. Which of the following would *not* be considered an isotope?
   a. ^2H (Deuterium)
   b. ^3H (Tritium)
   c. ^{12}C
   d. ^{14}C
   e. ^{32}P

7. Hydrogen bonds and hydrophilic interactions are
   a. weak chemical bonds that hold together atoms within a molecule.
   b. strong chemical bonds that hold together atoms within a molecule.
   c. weak chemical bonds that hold separate molecules together.
   d. strong chemical bonds that hold separate molecules together.
   e. interactions that hold protons and neutrons together within the atomic nucleus.

8. The main difference between an acid and a base is that
   a. bases are polar molecules and acids are not.
   b. acids are polar molecules and bases are not.
   c. bases donate hydrogen ions in water, whereas acids accept hydrogen ions.
   d. acids donate hydrogen ions in water, whereas bases accept hydrogen ions.
   e. one can act as a buffer, whereas the other cannot.

9. A pH value of 0 represents the lowest possible concentration of free hydrogen ions in a solution. (True or False)

10. The role of chemical buffers in living systems is to maintain the hydrogen ion concentration within narrow limits favorable for the internal environment of cells. (True or False)

11. A solution with pH = 5 is _____ than a solution with pH = 7.
    a. 2 times more basic
    b. 10 times more basic

c. 2 times more acidic

d. 10 times more acidic

e. 100 times more acidic

12. The molecule adenosine triphosphate (ATP) is most closely related to which class of macromolecules?

a. nucleic acids

b. fats

c. carbohydrates

d. proteins

e. membranes

13. Which of the following represents a polymer of amino acids?

a. ATP

b. fats

c. carbohydrates

d. proteins

e. membranes

14. In a saturated fat, you would expect to find all of the following *except*

a. single-bonded carbon atoms.

b. double-bonded carbon atoms.

c. hydrogen atoms.

d. nonpolar bonds.

e. covalent bonds.

15. Match each term with the best description.

__ phospholipid

__ carbohydrate

__ hydrophobic

__ acidity

__ ionic bond

a. major component of biological membranes

b. a nonpolar compound's interaction with water

c. an attraction of oppositely charged atoms

d. the cause of tartness in lemon juice

e. a polysaccharide

16. Which of the following would *not* be considered a type of sterol?

a. cholesterol

b. estrogen

c. testosterone

d. cellulose

e. vitamin D

Conceptual Understanding

1. In which of the following molecules would you expect to find electrons shared between the atoms that make up the molecule?

a. water

b. hydrogen gas

c. methane

d. hydrogen sulfide

e. all of the above

2. The strongest chemical linkages are found in molecules in which atoms are held together by _____ bonds.

a. noncovalent

b. covalent

c. ionic

d. hydrogen

e. hydrophilic

3. A compound capable of forming hydrogen bonds with water

a. is probably held together by noncovalent bonds.

b. contains at least some polar covalent bonds.

c. should act as a good buffer for acids and bases.

d. does not dissolve well in water.

e. is hydrophobic.

4. Salt and sugar both dissolve well in water, but for different reasons. In the case of salt, water molecules

a. form ionic bonds with the positively and negatively charged ions.

b. form nonpolar covalent bonds with the positively charged ions only.

c. surround the separated ions because of their partial electrical charge.

d. share electrons with the ions to make polar covalent bonds.

e. none of the above

5. Chemical reactions always begin and end with the same number and type of atoms, although those atoms are usually rearranged during the reaction. (True or False)

6. What will happen if you add a mild acid to a solution that is buffered at $pH = 7$?

a. The pH will drop well below 7.

b. The pH will rise well above 7.

c. The pH will not change much.

d. Ionic bonds will form between the acid and the buffer.

e. Covalent bonds will form between the acid and the buffer.

7. Which of the following is *not* an example of a compound that would fall into the category of monomers?

a. monosaccharides

b. DNA

c. amino acids

d. nucleotides

e. none of the above

8. Proteins, DNA, and starch are all
   a. carbon-based compounds.
   b. polymers.
   c. held together mainly by covalent bonds.
   d. made of atoms that contain protons and electrons.
   e. all of the above

9. A biological polymer is known to contain alanine, tyrosine, and lysine. This polymer is
   a. DNA.
   b. a strong base.
   c. a phospholipid.
   d. a protein.
   e. held together by hydrogen bonds.

10. Glucose is to cellulose as guanine is to
    a. DNA.
    b. starch.
    c. protein.
    d. phospholipids.
    e. fatty acids.

11. Consider the protein depicted in Figure 5.17. Assume that the protein is made exclusively from hydrophobic amino acids. This protein would also be considered
    a. polar.
    b. nonpolar.
    c. neutral.
    d. a buffer.
    e. none of the above

12. Which of the following components would *not* be found in a typical phospholipid bilayer?
    a. cholesterol
    b. glycerol
    c. fatty acids
    d. thymine
    e. carbohydrates

13. In a phospholipid bilayer, the hydrophobic tails of the phospholipid molecules are pointed outward so that they are in maximum contact with water. (True or False)

RELATED ACTIVITIES

- How have fast food and "junk" food manufacturers changed their products as a result of increased awareness of trans fats and the link to heart disease? Where are trans fats typically found in food products? Find an adult who has had food previously made with trans fat that is now made with no trans fat, and ask him or her what the difference is in taste. Which food is better tasting?
- Make a list of foods you ate today and identify the basic classes of molecules in each. Are all of the classes rep-

resented? How does this nutritional analysis relate to the concepts of biological chemistry discussed in the chapter?

- Refer to the Biology in the News box at the end of the chapter, "Do Trans Fats Cause Depression?" Write an argument for and against the point in this article. In your argument, make sure you include a discussion of the structure of cis and trans fats, and how and why they may or may not cause depression. Find a partner and determine who will argue for each side.

ANSWERS AND EXPLANATIONS

Factual Knowledge

1. d. Recall that hydrogen does not typically contain neutrons or have multiple electrons orbiting the nucleus. Also, ionic bonds are formed only between elements that can easily donate or accept electrons. All three of these elements are common to living organisms. For more information, see Section 5.1, Matter, Elements, and Atomic Structure.
2. b. Electrons have a negative electric charge and therefore would repel each other. For more information, see Section 5.1, Matter, Elements, and Atomic Structure.
3. c. Recall that ionic bonds are caused by the attraction of oppositely charged ions. Covalent bonds result from the sharing of electrons between atoms. For more information, see Section 5.2, *Covalent bonds form by electron sharing between atoms.*
4. a. Fixed ratios of specific atoms yield molecules, such as water and methane. For more information, see Section 5.2, Covalent and Ionic Bonds.
5. e. Water is made of hydrogen and oxygen, which are abundant elements. It is also a key component of many biological processes. Because water is polar, it will form hydrogen bonds with other polar molecules but not with nonpolar compounds. For more information, see Section 5.3, The Special Properties of Water, *Water is a polar molecule,* and *Water is a solvent for charged or polar substances.*
6. c. Recall that the atomic mass number of carbon is 12. Therefore, C-12 would not be considered an isotope, whereas C-14, with two extra neutrons, would. For more information, see Section 5.1, *Some elements exist in variant forms called isotopes.*
7. c. These bonds form between molecules, or within specific parts of molecules, to help stabilize structure. For more information, see Section 5.3, *Water is a polar molecule.*
8. d. Acids donate hydrogen ions to lower the pH of a solution, whereas bases accept hydrogen ions to raise

the pH. For more information, see Section 5.5, The pH Scale.

9. False. A pH of 0 represents the highest possible concentration of hydrogen ions and consequently the highest acidity. For more information, see Section 5.5, The pH Scale, and Figure 5.11.

10. True. Buffers are substances that can donate or accept hydrogen ions as necessary to maintain a stable pH. For more information, see Section 5.5, The pH Scale.

11. e. Recall that the pH scale is logarithmic, which means that each step on the scale represents a 10-fold increase or decrease in hydrogen ion concentration. A solution with a pH of 7 is neutral, whereas a solution with a pH of 5 is $10 \times 10 = 100$ times more acidic in comparison. Refer to Figure 5.11 for a description of the pH scale. For more information, see Section 5.5, The pH Scale, and Figure 5.11.

12. a. The "A" in ATP stands for "adenosine," a modified adenine nucleotide. For more information, see Section 5.10, Nucleotides and Nucleic Acids.

13. d. Polymers are repeating units of monomers. Amino acids are the monomers used to make proteins. For more information, see Section 5.8, Proteins.

14. b. "Saturated" means that the maximum number of hydrogen atoms are attached to each carbon, eliminating the possibility of double bonds between carbon atoms. For more information, see Section 5.9, Lipids, and Figure 5.18.

15. a. phospholipid. For more information, see Section 5.9, *Phospholipids are important components of cell membranes*.
    e. carbohydrate. For more information, see Section 5.7, Carbohydrates.
    b. hydrophobic. For more information, see Section 5.9, *Phospholipids are important components of cell membranes*.
    d. acidity. For more information, see Section 5.5, The pH Scale.
    c. ionic bond. For more information, see Section 5.2, *Ionic bonds form between atoms of opposite charge*.

16. d. Recall that sterols are a special class of lipid with many important biological functions. Cellulose, however, is a type of carbohydrate. For more information, see Section 5.9, *Sterols play vital roles in a variety of life processes*.

Conceptual Understanding

1. e. All of these are covalently bonded molecules. Electrons are thus shared between the atoms within the molecule. For more information, see Section 5.2, *Covalent bonds form by electron sharing between atoms*, and Figure 5.4, Covalent Bonds and Electron Shells.

2. b. Because electrons are actually shared between atoms, covalent bonds are the strongest types of chemical bonds. For more information, see Section 5.2, *Covalent bonds form by electron sharing between atoms*, and Figure 5.4, Covalent Bonds and Electron Shells.

3. b. Hydrogen bonds form between water molecules and the partially charged portions of other polar covalent molecules. For more information, see Section 5.3, *Water is a polar molecule*, and Figure 5.9, Hydrogen Bonding between Water Molecules.

4. c. The electrical charge present in salt ions attracts water molecules, which then surround them, keeping them separate from their electrically charged counterparts. For more information, see Section 5.3, *Water is a polar molecule* and *Water is a solvent for charged or polar substances*, and Figure 5.8, Charged Substances Dissolve in Water to Form Solutions.

5. True. Although new products are made in chemical reactions, the equation must always balance in terms of number and type of atoms involved. That is, matter can be neither created nor destroyed, only rearranged. For more information, see Section 5.4, Chemical Reactions.

6. c. Recall that the function of a buffer is to maintain the pH within a narrow range. For more information, see Section 5.5, The pH Scale.

7. b. DNA is a polymer made from nucleotide monomers. For more information, see Section 5.6, The Chemical Building Blocks of Life, *and* Section 5.10, Nucleotides and Nucleic Acids.

8. e. All of these statements are true. For more information, see Section 5.10, Nucleotides and Nucleic Acids; Section 5.8, Proteins; and Section 5.7, Carbohydrates. For background information, see Section 5.6, The Chemical Building Blocks of Life, *and* Section 5.2, *Covalent bonds form by electron sharing between atoms*.

9. d. Alanine, tyrosine, and lysine are all amino acids used to make proteins. For more information, see Section 5.8, *Proteins are built from amino acids, and* Figure 5.15b, The Structure and Diversity of Amino Acids.

10. a. Glucose is a building block used to make cellulose. Guanine is one of several building blocks used to assemble molecules of DNA. For more information, see Section 5.7, Carbohydrates, and Section 5.10, Nucleotides and Nucleic Acids.

11. b. Recall that hydrophobic amino acids are also nonpolar. As a result, this protein would be insoluble in water. For more information, see Section 5.8, *Proteins are built from amino acids*.

12. d. Recall that the nucleic acid thymine is one of the building blocks of DNA. You would not typically find

guanine associated with a phospholipid bilayer. For more information, see Section 5.9, *Phospholipids are important components of cell membranes*, and Figure 5.20, Membranes Contain Double Sheets of Phospholipids.

13. False. Just the opposite: the tails are pointed inward to avoid water. See the diagram in Figure 5.20. For more information, see Section 5.9, *Phospholipids are important components of cell membranes.*

CHAPTER 6 | Cell Structure and Internal Compartments

GETTING STARTED

Below are a few questions to consider before reading Chapter 6. These questions will help guide your exploration and assist you in identifying some of the key concepts presented in this chapter.

1. How many different types of cells are there in the human body?

2. What is the fluid mosaic model?

3. What are the "little organs" of the cell called?

4. If the cell were considered a tiny factory, which structure would serve as the shipping department? Which structure would serve as the power plant?

5. What are the three main components of the cytoskeleton, and what are their functions?

6. In a cell that is capable of swimming through water, what tiny structures would be analogous to the oars of a rowboat?

A GUIDE TO THE READING

The following concepts typically give students the most difficulty when exploring the content in Chapter 6 for the first time. For each concept, one or more references have been identified that may help you gain a better understanding of these potentially problematic areas.

Fluid Mosaic Model

The plasma membrane serves as the outer boundary of the cell. As we saw in Chapter 5, membranes are composed primarily of phospholipids arranged in a bilayer. Unfortunately, this bilayer arrangement restricts the movement of many molecules across the membrane. To import nutrients, export wastes, and communicate with its surroundings through the use of signaling molecules, the cell employs the services of special membrane proteins that are embedded in the phospholipid bilayer. These proteins perform specialized functions, such as the transport of larger molecules across the membrane. It is important to note that the proteins embedded in the plasma membrane typically are free to move laterally within the plane of the membrane. Because the membrane is composed of a mosaic of protein and phospholipids and the proteins are free to move fluidly, the plasma membrane of cells is often referred to as a "fluid mosaic." This characteristic is important for the cell's ability to move and to communicate with its environment.

For more information on this concept, be sure to focus on

- Section 6.2, The Plasma Membrane
- Figure 6.7, The Many Functions of Membrane Proteins

Cells Come in Different Sizes

Just as there is tremendous diversity in the types of cells found in the human body, so too there is a wide-ranging disparity in the sizes of cells. From the smallest of prokaryotes (less than 1 micrometer in size) all the way up to the largest of eukaryotic cells (greater than 1 meter in length), the size of cells is typically a function of how efficiently the cell is able to transport nutrients and waste materials throughout the cytoplasm. Eukaryotic cells, with an interworking series of organelles and cytoskeletal structures, are capable of actively transporting nutrients and wastes to their destination inside the cell. As a result, eukaryotic cells can be, and typically are, much larger than their prokaryotic counterparts (typically up to 1,000 times the volume).

For more information on this concept, be sure to focus on

- Section 6.3, Prokaryotic and Eukaryotic Cells
- Figure 6.8, Prokaryotic and Eukaryotic Cells Compared

Movement between Compartments

The cytosol of the eukaryotic cell contains numerous organelles that work together to carry out the activities of the cell. These activities include the production and export of proteins. This process begins at the "factory floor" of the cell, the endoplasmic reticulum (ER). Tiny structures called "ribosomes" attach to the surface of the ER function to produce proteins directly inside the membrane-enclosed ER (the lumen). To complete their journey, these proteins must escape the confines of the ER for processing and sorting in a different organelle—the Golgi apparatus. But how do they get there? The answer is through the use of tiny, membrane-bound transport structures called "transport vesicles." Transport vesicles are produced when parts of the ER begin to "bud" off from the organelle. Contents of the lumen of the ER, including the proteins destined for the Golgi apparatus, become trapped inside these buds, which finally break off, forming a transport vesicle. The transport vesicle then makes its way through the cytosol until it fuses with a nearby Golgi apparatus, dumping the trapped contents (including the protein) inside the Golgi. Once the proteins are processed inside the Golgi, they can be carried by another transport vesicle to their final destination. The budding and fusion of transport vesicles is made possible by the fluid nature of the bounding membranes. Because the inner portion of the phospholipid bilayers is hydrophobic, they will have a tendency to fuse together, just as two bubbles floating in the air or two droplets of oil floating on the surface of water might fuse together.

For more information on this concept, be sure to focus on

- Section 6.4, *The endoplasmic reticulum manufactures lipids and some proteins*
- Section 6.4, *Transport vesicles move materials*
- Section 6.4, *The Golgi apparatus sorts and ships macromolecules*
- Figure 6.11, Cellular Materials Are Dispatched to a Wide Variety of Destinations via Vesicles
- Figure 6.12, The Golgi Apparatus Routes Proteins and Lipids to Their Final Destinations

Cell Motility

The cytoskeleton of the cell provides both structure and movement to a cell. It consists of three distinct types of filaments: microtubules, intermediate filaments, and microfilaments. Microtubules are the largest of these structures. In eukaryotes, they facilitate cell motility in two ways. The first way involves the creation of a tiny system of "tracks," analogous to a subway system, starting at the center of the cell

(adjacent to the nucleus) and branching outward toward the plasma membrane. These tiny tracks are used by structures such as transport vesicles to crawl along to reach their final destination. Microtubules also provide the rigidity and flexibility necessary to produce tiny rowing (cilia) or beating (flagella) structures that serve to propel cells through a fluid medium.

In contrast, microfilaments are the smallest of the cytoskeletal structures. In eukaryotes, they facilitate cell movement by constantly adjusting their length and rearranging their alignment at the cell surface. The growth and realignment of the microfilaments causes small protrusions, termed "pseudopodia," to emerge from the cell, allowing the cell to grab hold and crawl along a surface. Intermediate filaments are not believed to serve a motility function in cells, but rather provide the cell with structural support.

For more information on this concept, be sure to focus on

- Section 6.5, The Cytoskeleton
- Figure 6.17, An Overview of the Cytoskeletal System
- Figure 6.18, The Structure of Microtubules, Intermediate Filaments, and Microfilaments
- Figure 6.19, Microfilaments Drive Some Types of Whole Cell Movement
- Figure 6.20, Cilia and Flagella Generate Movement

TYING IT ALL TOGETHER

Several concepts presented in this chapter build on those presented in previous chapters and will also be revisited and discussed in greater detail in subsequent chapters, including

Prokaryotes vs. Eukaryotes

- Chapter 3—Section 3.1, The Dawn of Eukarya
- Chapter 3—Section 3.2, Protista: The First Eukaryotes

Atomic Composition of Life

- Chapter 5—Section 5.1, Matter, Elements, and Atomic Structure

Biological Membranes

- Chapter 5—Section 5.9, Lipids

Cell Signaling and Communication

- Chapter 7—Section 7.6, Cell Signaling

Energy in Living Systems

- Chapter 8—Section 8.1, The Role of Energy in Living Systems

- Chapter 9—Section 9.4, Cellular Respiration: Breaking Down Molecules for Energy

Photosynthesis

- Chapter 9—Section 9.3, Photosynthesis: Capturing Energy from Sunlight

DNA

- Chapter 14—Section 14.1, An Overview of DNA and Genes

Protein Synthesis

- Chapter 15—Section 15.5, Translation: Information Flow from mRNA to Protein

PRACTICE QUESTIONS

Factual Knowledge

1. The simplest way to distinguish a prokaryotic from a eukaryotic cell is to check for
   a. a plasma membrane.
   b. a nucleus.
   c. DNA.
   d. proteins.
   e. whether the cell is a single-celled organism.

2. Which of the following structures would you expect to find in a bacterium?
   a. nucleus
   b. plasma membrane
   c. Golgi apparatus
   d. lysosome
   e. endoplasmic reticulum (ER)

3. Which organelle is surrounded by a double phospholipid bilayer with many pores?
   a. nuclear envelope
   b. plasma membrane
   c. Golgi apparatus
   d. mitochondrion
   e. all of the above

4. Which of the following structures is considered an organelle?
   a. microtubule
   b. cytosol
   c. vacuole
   d. plasma membrane
   e. DNA

5. Proteins that are destined for insertion into a membrane are one product of the rough ER. (True or False)

6. The Golgi apparatus is one destination for substances that have been packaged into transport vesicles by rough ER. (True or False)

7. Lysosomes are specialized vesicles in _____ that contain digestive enzymes for the breakdown of food. A related organelle known as a "vacuole," which is found in _____, also contains enzymes but in addition may act as a storage organelle for nutrients or water.
   a. animals; plants
   b. plants; animals
   c. fungi; animals
   d. plants; fungi
   e. animals; fungi

8. Both chloroplasts and mitochondria
   a. have multiple membranes.
   b. have highly structured innermost membranes.
   c. are found only in eukaryotic cells.
   d. are involved in cellular energy processing.
   e. all of the above

9. In which of the following cell organelles would you expect to find the biochemical reactions that harness energy from the breakdown of sugar molecules to synthesize large amounts of adenosine triphosphate (ATP)?
   a. lysosome
   b. vesicles
   c. chloroplast
   d. mitochondrion
   e. plasma membrane

10. Microtubules, motor proteins, and actin filaments are all part of the
    a. mechanism of photosynthesis that occurs in chloroplasts.
    b. rough ER in prokaryotic cells.
    c. cytoskeleton of eukaryotic cells.
    d. process that moves small molecules across cell membranes.
    e. none of the above

11. Which of the following macromolecules are found in the plasma membrane?
    a. lipids only
    b. lipids and proteins
    c. lipids, proteins, and nucleic acids
    d. proteins and nucleic acids
    e. proteins only

12. The vesicles that bud off of the ER contain free-floating molecules from the lumen as well as molecules embedded in the membrane. (True or False)

13. Which of the following organelles would likely have the lowest pH?
    a. nucleus
    b. mitochondrion
    c. lysosome
    d. endoplasmic reticulum (ER)
    e. more than one of the above

Conceptual Understanding

1. In terms of basic structure, elephant cells and oak tree cells both
   a. are prokaryotes.
   b. have chloroplasts.
   c. have a cell wall.
   d. have mitochondria.
   e. all of the above

2. In which of the following cell types would you see more cellular detail using an electron microscope than you would with a light microscope?
   a. animal
   b. plant
   c. bacterial
   d. protist
   e. all of the above

3. Which of the following statements about prokaryotes is *false*?
   a. Just like eukaryotes, prokaryotes have a plasma membrane.
   b. Prokaryotic cells concentrate important materials for the cell's survival.
   c. The substance known as "cytosol" is found within the bacterial nucleus.
   d. Prokaryotic cells are much smaller than most eukaryotic cells.
   e. Ancient prokaryotes may have given rise to some eukaryotic organelles.

4. You are asked to examine a cell using a powerful light microscope. The image you see has a clearly defined nucleus and mitochondria. It also has a large vacuole and mitochondria. From what group of organisms did this cell most likely come?
   a. bacteria
   b. fungi
   c. plants
   d. animals
   e. cannot tell

5. According to the fluid mosaic model of cell membranes,
   a. proteins are rigidly fixed in the phospholipid bilayer.
   b. the most common types of molecules in the membrane are proteins.
   c. basic membrane structure results from the ways in which proteins interact with water.
   d. the membrane is a highly mobile mixture of phospholipids and proteins.
   e. the unique properties of cell types are determined by their phospholipids.

6. The best way to identify a cell as either prokaryotic or eukaryotic is to determine whether
   a. it came from a single-celled or multicellular organism.
   b. it contains DNA enclosed within a membrane-bound compartment.
   c. it has a cell wall.
   d. it has cytosol.
   e. all of the above

7. Eukaryotic cells are more efficient than prokaryotes because their internal compartmentalization
   a. makes each compartment nutritionally independent of all others.
   b. allows for specialization through the subdivision of particular tasks.
   c. permits the unregulated flow of materials around the cell.
   d. eliminates the need for communication with the external environment.
   e. reduces overall cell size.

8. In a microscope, you observe an organelle that consists of a series of interconnected tubes and appears to be connected to the outer membrane of the nucleus in a eukaryotic cell. What is the likely identity of this organelle?
   a. mitochondrion
   b. lysosome
   c. Golgi apparatus
   d. vacuole
   e. endoplasmic reticulum (ER)

9. Consider a protein that is destined to be embedded in the plasma membrane. This protein must first be assembled in the _____ and then transported to the _____ for processing and sorting.
   a. smooth ER; rough ER
   b. mitochondria; Golgi apparatus
   c. Golgi apparatus; transport vesicle
   d. rough ER; Golgi apparatus
   e. lysosome; smooth ER

10. How do actin and tubulin proteins relate to eukaryotic cell structure and function?
    a. They are both embedded proteins in plasma membranes.
    b. They are both components in the reactions of photosynthesis.

c. They are both participants in the production of large amounts of ATP.
d. They both facilitate cell motility.
e. none of the above

11. Which of the following structures would you expect to find associated with motor proteins?
   a. nuclear pores
   b. ribosomes
   c. vesicles
   d. plasma membranes
   e. all of the above

12. Which of the following structures produces motion through the action of filaments that "slide" or "walk" parallel to each other?
   a. eukaryotic cilia
   b. eukaryotic flagella
   c. prokaryotic flagella
   d. a and b only
   e. b and c only

13. Which of the following statements regarding Tay-Sachs disease is inaccurate?
   a. Individuals with Tay-Sachs lack a particular lysosomal enzyme.
   b. Individuals with Tay-Sachs accumulate excess lipid in their nerve cells.
   c. Restricting marriage to a small community can increase the probability that a child may be born with a disease such as Tay-Sachs.
   d. The bacterium that causes Tay-Sachs is effective at burrowing deep into the host's cells to evade the immune system.
   e. all of the above

Related Activities

- Make a Venn diagram of the similarities and differences between prokaryotic and eukaryotic cells. Be sure to include not only basic aspects of cell structure, such as the specific types of macromolecules that are present, but also the presence or absence of specialized features such as organelles and biochemical processes.
- Refer to an online encyclopedia and look up information on microscopes. In addition to contrast and magnification, another important aspect of a microscope's functioning involves the concept of resolution. Explain what resolution means in the context of microscopy, and explain how it relates to our ability to define structures present within cells. Why would this be considered important in your daily life? Determine what (if any) organelles can be seen under a compound light microscope.
- Make a list of all major buildings and traffic routes in an average town or city (for example, a highway, the police station, city hall). On the same paper, make a list of different parts of a cell, and then determine which part of the town or city would be each component of the cell. Describe how your town or city would work if one of your cell components was not working properly. Draw a sketch of what your city would look like, then label each part of your town or city using cell component names and show how each corresponds to the town or city.

ANSWERS AND EXPLANATIONS

Factual Knowledge

1. b. All cells have plasma membranes, DNA, and proteins. There are single-celled representatives from both prokaryotes and eukaryotes. Of the choices given, the most reliable way to tell the difference between cell types is the presence of a nucleus, which is found only in eukaryotes. For more information, see Section 6.3, Prokaryotic and Eukaryotic Cells.

2. b. All of the structures listed are eukaryotic organelles, but plasma membrane is the only listed structure that is *also* in a prokaryotic organism such as a bacterium. For more information, see Section 6.3, Prokaryotic and Eukaryotic Cells, and Figure 6.8, Prokaryotic and Eukaryotic Cells Compared.

3. a. The pores in the nuclear envelope permit relatively large molecules such as proteins and RNA to enter or exit this organelle. For more information, see Section 6.4, *The nucleus houses genetic material*, and Figure 6.9, The Nucleus Contains DNA, the Genetic Material of the Cell.

4. c. Vacuoles are found in plants. They serve as a type of storage compartment in the cell. The other structures listed would not be considered true organelles. For more information, see Section 6.4, *Lysosomes and vacuoles disassemble macromolecules*.

5. True. Proteins destined for use in the cytosol are made by free-floating ribosomes rather than by ribosomes bound to the rough ER. For more information, see Section 6.4, *The endoplasmic reticulum manufactures lipids and some proteins*, and Figure 6.10, Some Types of Lipids and Proteins Are Made in the Endoplasmic Reticulum.

6. True. Rough ER produces transport vesicles, some of which find their way to the Golgi apparatus, where additional processing of their contents occurs before those contents are directed to their final destinations. For more information, see Section 6.4, *The Golgi apparatus sorts and ships macromolecules* and *The endoplasmic reticulum manufactures lipids and some proteins*.

7. a. Lysosomes are present in animals because of animals' special need for internal digestion of macromolecules. Plants store enzymes, nutrients, and water in their large vacuoles, structures that are not present in animals. For more information, see Section 6.4, *Lysosomes and vacuoles disassemble macromolecules*.

8. e. For more information, see Figures 6.15 and 6.16 in your textbook. Also, see Section 6.4, *Mitochondria power the cell*, and Section 6.4, *Chloroplasts capture energy from sunlight*.

9. d. The mitochondria are the powerhouse of the cell. They are where the bulk of the reactions that manufacture ATP take place. For more information, see Section 6.4, *Mitochondria power the cell*.

10. c. The cytoskeleton supports the cell and allows for movement of the entire cell and also within the cell. All three molecules listed are part of the cytoskeleton. For more information, see Section 6.5, The Cytoskeleton. See also Section 6.5, *Microtubules support movement inside the cell*, and Figure 6.18, The Structure of Microtubules, Intermediate Filaments, and Microfilaments.

11. b. The chapter refers to the mixture of lipids and proteins creating a "fluid mosaic." For more information, see Section 6.2, The Plasma Membrane, and Figure 6.7, The Many Functions of Membrane Proteins.

12. True. For more information, see Figure 6.11 for a diagram of how this occurs. See also Section 6.4, *The endoplasmic reticulum manufactures lipids and some proteins*.

13. c. Recall that the lysosome serves as the digestive center of the cell. It contains specialized enzymes that break down various macromolecules. These enzymes operate most efficiently in an acidic environment; therefore, the pH of the lysosomes is typically around 5, whereas the remainder of the organelles listed would have more neutral (pH 7) internal environments. For more information, see Section 6.4, *Lysosomes and vacuoles disassemble macromolecules*.

Conceptual Understanding

1. d. Because elephants and oak trees are both eukaryotes, they would both require the energy production functionality of mitochondria for survival. Recall that only plant cells have cell walls and chloroplasts. For more information, see Section 6.3, Prokaryotic and Eukaryotic Cells. See also Section 6.4, *Mitochondria power the cell*.

2. e. Electron microscopes have much greater magnifying power than light microscopes. This feature permits us to see more detail for whatever cell type we are viewing. For more information, see Section 6.1, *The microscope is a window into the life of a cell*.

3. c. Cytosol is found in prokaryotes but not in the nucleus, since this organelle would be absent. For more information, see Section 6.3, Prokaryotic and Eukaryotic Cells.

4. c. We can eliminate bacteria because they lack organelles completely. Among the organelles listed, the presence of a large vacuole implies that the cell is from a plant. For more information, see Section 6.1, *The microscope is a window into the life of a cell*. See also Section 6.3, Prokaryotic and Eukaryotic Cells.

5. d. The fluid aspect refers to the fact that both phospholipids and proteins are somewhat mobile within the bilayer. The mosaic refers to the patchwork distribution of proteins across the membrane. For more information, see Section 6.2, The Plasma Membrane.

6. b. The only definitive characteristic listed is the presence or absence of a nucleus (the membrane-bound compartment in which the cell stores DNA). For more information, see Section 6.3, Prokaryotic and Eukaryotic Cells.

7. b. Like specialized tasks in a corporation or society, the division of labor within eukaryotic cells greatly increases overall cellular efficiency. For more information, see Section 6.3, Prokaryotic and Eukaryotic Cells, and Section 6.4, Internal Compartments of Eukaryotic Cells.

8. e. The ER consists of a series of interconnected tubes scattered throughout the cytosol. The connection to the nucleus allows material produced in the nucleus to directly enter the ER through the nuclear pores. For more information, see Section 6.4, *The endoplasmic reticulum manufactures lipids and some proteins*.

9. d. Recall that proteins destined for the plasma membrane are produced by ribosomes found on the surface of the rough ER. Following assembly of the protein, it is then transported inside a vesicle to the Golgi apparatus, where it is processed, sorted, and sent off to its final destination (the plasma membrane). For more information, see Section 6.4, *The endoplasmic reticulum manufactures lipids and some proteins* and *The Golgi apparatus sorts and ships macromolecules*.

10. d. Both actin and tubulin are proteins that make up parts of the cytoskeleton. Recall that one of the prime functions of the cytoskeleton is to facilitate the process of cell movement. For more information, see Section 6.5, *Microtubules support movement inside the cell* and *Microfilaments are involved in cell movement*.

11. c. Transport vesicles are carried throughout the cell by way of motor proteins that typically crawl along microtubule "tracks." For more information, see Section 6.5, *Microtubules support movement inside the cell*.

12. d. Both eukaryotic cilia and flagella produce movement by the sliding action of parallel microtubules,

which is driven by motor proteins such as dynein. Prokaryotic flagella rely on a completely different mechanism. For more information, see Section 6.5, *Cilia and flagella enable whole cells to move.*

13. d. Recall that Tay-Sachs disease is caused by having a defective gene for a particular lysosomal enzyme that helps eliminate excess lipids from nerve cell membranes. *Listeria monocytogenes* is a bacterium that can burrow deep inside cells to evade the immune system and can cause severe food poisoning. See the Biology Matters box, "Organelles and Human Disease," in Chapter 6.

CHAPTER 7 | Cell Membranes, Transport, and Communication

GETTING STARTED

Below are a few questions to consider before reading Chapter 7. These questions will help guide your exploration and assist you in identifying some of the key concepts presented in this chapter.

1. Why are low-density lipoprotein (LDL) receptors so important for human health?

2. What properties of a phospholipid bilayer make it selectively permeable?

3. What percentage of the energy expended by a person at rest is used to fuel active transport across the plasma membrane?

4. What are the differences between the terms "hypotonic," "hypertonic," and "isotonic"?

5. How many gallons of water are contained in a typical person's body?

6. What mechanisms facilitate cellular "eating" and "drinking" activities?

7. What property of steroid hormones allows them to pass easily through the plasma membrane of target cells?

A GUIDE TO THE READING

The following concepts typically give students the most difficulty when exploring the content in Chapter 7 for the first time. For each concept, one or more references have been identified that may help you gain a better understanding of these potentially problematic areas.

Osmosis

As we saw in the chapter, the process of diffusion drives molecules to spread from areas of relatively high concentration to areas of lower concentration. In a similar fashion, when two solutions of different concentrations are separated by a selectively permeable membrane (which restricts the passage of the dissolved molecules), water molecules will exhibit directional movement. The movement of water molecules across a selectively permeable membrane is referred to as "osmosis." The direction in which water moves across the membrane is a function of the relative concentrations of the two solutions. As an example, consider Solution A, which has a relatively higher concentration of dissolved molecules than Solution B. In this example, Solution A would be considered "hypertonic" relative to Solution B, because it has a higher concentration of dissolved molecules. In contrast, Solution B, with a lower concentration of dissolved molecules, would be considered "hypotonic." The key to understanding these two terms is realizing that "hypertonic" and "hypotonic" can be used only when comparing one solution to another; a single solution cannot be hypertonic or hypotonic because there would be no companion solution to compare it to. In terms of osmosis, water molecules will always move across a membrane from the hypotonic solution toward the hypertonic solution. This can actually be thought of in terms of diffusion, because the hypertonic solution, which has a higher concentration of dissolved molecules taking up space, would actually have less room for water molecules. As a result, hypertonic solutions would have relatively fewer water molecules than hypotonic solutions, whereas hypotonic solutions would have relatively more water molecules. If you now think of osmosis as the diffusion of water, it is easy to see why water would always move from the hypotonic solution (relatively higher concentration of water) toward the hypertonic (relatively lower concentration of water) solution. In addition, solutions that have the same concentration can be described as being isotonic.

For more information on this concept, be sure to focus on

- Section 7.2, Osmosis
- Figure 7.5, Water Moves into and out of Cells by Osmosis

Cell Communication

Multicellular organisms require a reliable, efficient mechanism for facilitating communication among the cells of the body. This is most typically accomplished by the release of a signaling molecule (proteins, hormones, gasses) from one cell, which is then received by a target cell, which then responds in a highly specific fashion. The types of signaling molecules and their effects on their respective target cells vary greatly. One thing to understand about signaling molecules and their target cells is the fact that in order for a target cell to receive and respond to a chemical signal, it must possess a specific receptor protein for that particular signal. These receptor proteins typically are found on the surface of the target cell and work to transmit a signal into the cell when a signaling molecule binds to it. Other classes of receptor proteins may actually be located in the cytoplasm of the cell. In this case, the signaling molecule must be permeable across the plasma membrane in order to stimulate the target cell to respond in the desired manner. Another key concept to understand about signaling molecules is that their longevity depends on how far the molecule must travel to reach its intended target cell.

For more information on this concept, be sure to focus on

- Section 7.5, Cellular Connections
- Section 7.6, Cell Signaling
- Figure 7.12, Receptors for Signaling Molecules

TYING IT ALL TOGETHER

Several concepts presented in this chapter build on those presented in previous chapters and will also be revisited and discussed in greater detail in subsequent chapters, including

Kingdoms of Life

- Chapter 2—Section 2.1, *All of life on Earth can be sorted into three distinct domains*

Macromolecules

- Chapter 5—Section 5.6, The Chemical Building Blocks of Life

The Plasma Membrane

- Chapter 6—Section 6.2, The Plasma Membrane

The Cytoskeleton

- Chapter 6—Section 6.5, The Cytoskeleton

Genes and Proteins

- Chapter 15—Section 15.2, How Genes Guide the Manufacture of Proteins

Hormones

- Chapter 29—Section 29.1, How Hormones Work

PRACTICE QUESTIONS

Factual Knowledge

1. Which of the following macromolecules are found in the plasma membrane?
   a. lipids only
   b. proteins only
   c. lipids and proteins
   d. lipids and nucleic acids
   e. nucleic acids only

2. Which of the following plasma membrane components is responsible for allowing the passage of charged molecules, such as sodium or chloride, against a concentration gradient?
   a. phospholipid bilayer
   b. channel proteins
   c. gap junctions
   d. active carrier proteins
   e. receptor proteins

3. The process of dissolving a powdered drink in a glass of water demonstrates the principle of
   a. active transport.
   b. diffusion.
   c. selective permeability.
   d. fluid mosaic.
   e. none of the above

4. Osmosis is the
   a. movement of molecules down a concentration gradient.
   b. movement of molecules through a channel protein.
   c. movement of large, biologically important molecules across a membrane.
   d. movement of water across a membrane.

e. active transport of molecules against a concentration gradient.

5. A hypertonic solution
   a. has a lower concentration relative to another solution.
   b. has a higher concentration relative to another solution.
   c. has a concentration equal to another solution.
   d. requires the input of energy.
   e. always loses water.

6. The process of phagocytosis is considered to be a type of
   a. endocytosis.
   b. exocytosis.
   c. pinocytosis.
   d. passive transport.
   e. none of the above

7. Which of the following would *not* be considered a type of cell junction?
   a. plasmodesmata
   b. extracellular matrix
   c. gap junctions
   d. tight junctions
   e. anchoring junctions

8. Two key principles important for the evolution of large multicellular organisms are
   a. prokaryotic cell structure and cell specialization.
   b. cell specialization and communication among cells.
   c. communication among cells and simple nerve reflexes.
   d. simple nerve reflexes and cell specialization.
   e. prokaryotic cell structure and simple nerve reflexes.

9. The class of signaling molecules called "steroid hormones"
   a. exert their effects at a location away from the site of production.
   b. are hydrophilic and cannot penetrate the plasma membrane.
   c. bind exclusively to cell surface receptors.
   d. never enter the blood of humans.
   e. all of the above

10. A signal transduction pathway involves
    a. a signal binding to a cell surface receptor, which then transmits the signal inside the cell.
    b. a signal binding to a receptor located in the cytoplasm of the cell.
    c. a signal binding directly to the plasma membrane.
    d. a signal that never reaches the intended target cell.
    e. none of the above

11. Which of the following is *not* a type of signaling molecule?
    a. neurotransmitters
    b. human growth hormone (hGH)
    c. insulin
    d. sulfur dioxide
    e. testosterone

12. Match each term with the best description.
    __ osmoregulation
    __ phagocytosis
    __ plasmodesmata
    __ GLUT proteins
    __ hormones
    a. specialized glucose carriers
    b. small water-transport channels
    c. long-range signaling molecules
    d. cell eating
    e. process of maintaining water balance

Conceptual Understanding

1. Consider a relatively small, complex biochemical, such as an amino acid. If this molecule passes through the plasma membrane to gain entry into the cell, it will most likely pass
   a. directly through the phospholipid bilayer.
   b. through an active carrier protein.
   c. through a channel protein.
   d. through a tight junction.
   e. though an extracellular matrix.

2. Consider the small, biologically important molecules water, oxygen, and carbon dioxide. If these molecules enter or exit the cell, they will most likely pass
   a. directly through the phospholipid bilayer.
   b. through an active carrier protein.
   c. through a channel protein.
   d. through a tight junction.
   e. through a gap junction.

3. Consider two solutions (A + B) containing a mixture of water and dissolved proteins that are separated by a membrane permeable only to water. If Solution A has a relatively higher concentration of proteins,
   a. Solution A would be considered hypertonic.
   b. Solution B would be considered hypotonic.
   c. neither solution would be considered isotonic.
   d. all of the above
   e. none of the above

4. Consider two solutions (A + B) containing a mixture of water and dissolved proteins that are separated by a membrane permeable only to water. If Solution A has a relatively higher concentration of proteins,
   a. proteins would move from Solution A to Solution B.

b. water would move from Solution A to Solution B.
c. proteins would move from Solution B to Solution A.
d. water would move from Solution B to Solution A.
e. both water and proteins would have no net movement across the membrane.

5. In our liver cells, receptors capable of recognizing LDL cholesterol particles bind to these particles and stimulate their uptake. The process by which these particles are taken into the cell is called
   a. exocystosis.
   b. phagocytosis.
   c. pinocytosis.
   d. receptor-mediated endocytosis.
   e. reverse osmosis.

6. Which of the following cell junction types would *not* allow for the passage of small signaling molecules?
   a. plasmodesmata
   b. tight junctions
   c. anchoring junctions
   d. gap junctions
   e. none of the above

7. Which of the following statements regarding signaling molecules is *false*?
   a. Signaling molecules are widely used among multicellular organisms.
   b. All signaling molecules have the same effects on their target cells.
   c. All signaling molecules require the action of specific receptor molecules present in the target cell.
   d. Signaling molecules have different longevities depending on their activity and function.
   e. Signaling molecules have a direct effect on their target cells.

8. Which of the following types of transport and proteins would be required to move a large, charged molecule against its concentration gradient?
   a. passive transport, channel proteins
   b. passive transport, carrier proteins
   c. active transport, channel proteins
   d. active transport, carrier proteins
   e. none of the above

9. Which of the following statements regarding a typical hydrophilic signaling molecule would be true?
   a. Its receptor would be located in the cytosol of the target cell.
   b. Its receptor would be located on the surface of the cell.
   c. After entering the cell, it would travel directly to the nucleus.

d. Because it can enter the cell, it directly affects specific cellular processes.
e. It is most likely a steroid.

10. A cell is known to respond to a particular signaling molecule. Which of the following must be true of this cell?
    a. It is in the heart muscle.
    b. It responds to both nitric oxide and adrenaline.
    c. It is also the site of production for the signaling molecule.
    d. It contains the receptor for the signaling molecule.
    e. It is an animal cell.

11. A substance is brought to you that is a suspected steroid hormone. If this tentative identification is correct, what do you automatically know is true about this substance?
    a. It is likely a protein.
    b. It will likely have its cellular action very near its source of production.
    c. It will bind to and activate an intracellular receptor.
    d. It induces cell division.
    e. all of the above

12. A sudden release of adrenaline in a person undergoing a frightening experience will result in the release of stored fuels, such as the energy stored in fat. (True or False)

RELATED ACTIVITIES

- One of the classes of signaling molecules discussed in this chapter consists of the steroid hormones, which are derived from cholesterol produced by the body. Using your library or the Internet, perform a search to determine some of the other functions cholesterol plays in our bodies. Compose a one-page summary of your results; it should include a brief discussion of what medical conditions are associated with an overproduction of cholesterol.
- Drawing from the content presented in this chapter, search the Internet to answer the question of why some signaling molecules have their effects all over the body (for example, metabolism), whereas others have effects only at specific locations in the body (for example, development of sexual maturity). Using your results, compose a one-page essay describing additional examples of cell communication, besides metabolism and sexual maturation, that would fall into the categories of general or specific signaling effects.
- The Biology in the News box at the end of the chapter, " 'Good' Cholesterol May Lower Alzheimer's Risk,"

discusses the effect of high-density lipoprotein (HDL), or "good" cholesterol, on decreasing Alzheimer's risk. Research ways to increase the intake of HDL in your daily life and ways to decrease your LDL, or "bad" cholesterol. How else can making changes in your lifestyle to improve your HDL intake improve your overall health? Make a list of lifestyle changes and foods to eat or avoid to help increase the daily HDL intake.

ANSWERS AND EXPLANATIONS

Factual Knowledge

1. c. Recall that the plasma membrane consists of a phospholipid bilayer along with a series of proteins that perform specific functions for the cell, such as the channel and active or passive carrier proteins discussed in the chapter. For more information, see Section 7.1, The Plasma Membrane as Gate and Gatekeeper.

2. d. Recall that active carrier proteins are used by the cell to move molecules against a concentration gradient. These proteins require the expenditure of energy (adenosine triphosphate, or ATP) to accomplish this feat. For more information, see Section 7.3, *Active carrier proteins move materials against a concentration gradient.*

3. b. Recall that the passive (no energy required) transport of molecules down a concentration gradient illustrates the principle of diffusion. For more information, see Section 7.1, *In diffusion, substances move passively down a concentration gradient.*

4. d. Osmosis is the movement of water across a membrane. The direction of movement depends on the relative concentrations of the solutions on either side of the membrane. For more information, see Section 7.2, Osmosis.

5. b. Recall that the terms "hypotonic," "hypertonic," and "isotonic" can be used only when comparing two or more solutions. A hypertonic solution always has a relatively higher concentration. For more information, see Section 7.2, Osmosis.

6. a. Phagocytosis, also called "cell eating," involves the cell engulfing large particles. This process is a type of endocytosis. For more information, see Section 7.4, Endocytosis and Exocytosis, and Figure 7.10, Extracellular Substances Are Imported through Endocytosis.

7. b. Although the extracellular matrix plays a role in binding cells together, it does not contribute to the overall communication between cells, a primary function of cell junctions. For more information, see Section 7.5, Cellular Connections.

8. b. Cell communication and specialization are the two main organizing principles of multicellular organisms discussed in this chapter. For more information, see Section 7.5, Cellular Connections, and Section 7.6, Cell Signaling.

9. a. Steroid hormones are derived from cholesterol and can therefore pass through membranes to also bind with receptors in the cytosol. They travel all over the body, mainly by way of the blood, and have target cells that can be located a considerable distance from the source of hormone production. For more information, see Section 7.6, Cell Signaling.

10. a. Recall that when a signaling molecule binds to a plasma membrane receptor, the signal must be relayed to the cytoplasm. This occurs through a series of events that constitute a signal transduction pathway. For more information, see Section 7.6, Cell Signaling.

11. d. Sulfur dioxide is an environmental contaminant that contributes to the development of acid rain. Insulin, testosterone, and hGH are all hormones that are signaling molecules, and neurotransmitters are fast-acting signaling molecules. Refer to Figure 7.12 in your textbook for examples of receptors for signaling molecules. For more information, also see Section 7.6, Cell Signaling.

12. e. osmoregulation. For more information, see Section 7.2, Osmosis.
    d. phagocytosis. For more information, see Section 7.4, Endocytosis and Exocytosis, and Figure 7.10, Extracellular Substances Are Imported through Endocytosis.
    b. plasmodesmata. For more information, see Section 7.5, Cellular Connections, and Figure 7.11, Cells in Multicellular Organisms Are Interconnected in Various Ways.
    a. GLUT proteins. For more information, see Section 7.3, *Passive carrier proteins mediate facilitated diffusion.*
    c. hormones. For more information, see Section 7.6, Cell Signaling.

Conceptual Understanding

1. b. Recall that carrier proteins located in the plasma membrane facilitate the transport of larger biochemicals, such as amino acids, sugars, and small proteins. For more information, see Section 7.3, Facilitated Membrane Transport and *Active carrier proteins move materials against a concentration gradient.*

2. a. Recall that small molecules such as water, oxygen, and carbon dioxide can pass directly through the plasma membrane without the expenditure of energy. This makes the transport of these biologically important molecules easy and efficient for cells. For more

information, see Section 7.1, The Plasma Membrane as Gate and Gatekeeper.

3. d. The terms "hypertonic," "hypotonic," and "isotonic" are used to describe the relative concentrations of two solutions. In this case, choices "a," "b," and "c" are all accurate descriptions. For more information, see Section 7.2, Osmosis, and Figure 7.5, Water Moves into and out of Cells by Osmosis.

4. d. The terms "hypertonic," "hypotonic," and "isotonic" are used to describe the relative concentrations of two solutions. In this case, the process of osmosis would draw water from Solution B into Solution A across the membrane. Section 7.2, Osmosis, and Figure 7.5, Water Moves into and out of Cells by Osmosis.

5. d. The process of endocytosis is used by the cell to import external compounds. In receptor-mediated endocytosis, specialized receptor proteins are responsible for identifying and determining which substances, such as LDL cholesterol, enter the cell. For more information, see Section 7.4, Endocytosis and Exocytosis. See also Figure 7.9, Cell Contents Are Exported through Exocytosis, and Figure 7.10, Extracellular Substances Are Imported through Endocytosis.

6. b. Recall that tight junctions are responsible for holding cells together using strands of rigid proteins. Tight junctions form leak-proof barriers that prevent the passage of molecules. Refer to Figure 7.11 for additional information. For more information, see Section 7.5, Cellular Connections.

7. b. The activity and effect of a signaling molecule on its target cell are highly variable and depend on the nature of both the signal and the receptor involved in the response. For more information, see Section 7.5, Cellular Connections, and Section 7.6, Cell Signaling.

8. d. Active transport is the only type of transport that requires energy to help carry a molecule against its concentration gradient. A channel protein would not work to move a molecule against its concentration gradient and requires a carrier protein to move the molecule by using energy. For more information, see Section 7.3, Facilitated Membrane Transport, and Figure 7.6, The Plasma Membrane Controls What Enters and Leaves the Cell.

9. b. Hydrophilic molecules cannot cross membranes, so the way this signaling molecule would have its effect is by interacting with a cell surface receptor. For more information, see Section 7.6, Cell Signaling, and Figure 7.12, Receptors for Signaling Molecules.

10. d. Only target cells with appropriate receptors can respond to signaling molecules. For more information, see Section 7.6, Cell Signaling.

11. c. Because steroid hormones, such as progesterone, are derived from cholesterol and are hydrophobic, they pass directly through the plasma membrane and bind to and activate intracellular receptors. Once activated, these steroid-protein complexes stimulate the production of specific proteins that are necessary for the target cell's response. For more information, see Section 7.6, Cell Signaling.

12. True. This effect of adrenaline makes possible the conversion of fats into glucose for energy. For more information, see Section 7.6, Cell Signaling.

CHAPTER 8 | Energy, Metabolism, and Enzymes

GETTING STARTED

Below are a few questions to consider before reading Chapter 8. These questions will help guide your exploration and assist you in identifying some of the key concepts presented in this chapter.

1. How does using an electric blender to make a fruit smoothie demonstrate the principle of the first law of thermodynamics?

2. How does an aging tool shed demonstrate the principle of the second law of thermodynamics?

3. What is the primary energy source for all living organisms?

4. How is the activation energy for a chemical reaction similar to a safety match?

5. What are enzymes, and what function do they play in biochemical reactions?

6. How might metabolism shorten the life span of living organisms?

7. How many minutes of energetic dancing would be needed to burn the number of calories consumed from a large hamburger and fries?

A GUIDE TO THE READING

The following concepts typically give students the most difficulty when exploring the content in Chapter 8 for the first time. For each concept, one or more references have been identified that may help you gain a better understanding of these potentially problematic areas.

Laws of Thermodynamics

Thermodynamics is a dedicated field of physics that studies the dynamics of heat (from the Greek "thermo"). Heat is a form of energy that is involved in almost all chemical reactions. Certain universal laws govern how energy behaves in all chemical reactions; these are the laws of thermodynamics. As we saw in the chapter, the first law of thermodynamics dictates that energy cannot be created (or destroyed) but can only be transformed from one form to another. An excellent example of this is the gasoline used to power automobiles. Gasoline contains energy in a liquid form that is stable at room temperature. When gas is burned inside an engine, the energy is released. This energy takes on different forms: force (used to power the movement of the pistons, propelling the car forward) and heat. In this example, the amount of energy contained in the gasoline exactly equals the amount of energy utilized as force and lost as heat. In this and all cases, energy is not destroyed or created—it is merely transformed from one form (liquid gasoline) to another (force and heat). The second law of thermodynamics dictates that all systems (for example, a cell, a bedroom, a yard, the biosphere) always tend to become more disordered. An excellent example of this would be your dormitory room. Over time, the room will always become messier (disordered). The only way to reverse this trend is through the expenditure of energy (cleaning your room). Without this input of energy to restore order to the system, it will remain in a perpetual state of disorder. The use of energy in a cell is governed by these two laws of thermodynamics without exception. The multitude of chemical reactions that provide the cell's metabolism must always follow these rules.

For more information on this concept, be sure to focus on

- Section 8.1, *The laws of thermodynamics apply to living systems*
- Figure 8.2, The Second Law of Thermodynamics

Oxidation and Reduction Reactions

The process of electron transfer plays a crucial role in the multitude of chemical reactions in which cells use, capture, and store energy. The transfer of electrons between molecules will result in one molecule gaining electrons (reduction) and the other molecule losing electrons (oxidation). It is important to note that oxidation and reduction often occur as a paired process. Because of this close association, the textbook indicates these reactions are often called "oxidation-reduction," or "redox," reactions. The loss of electrons is called "oxidation" because when carbon atoms lose electrons through this process, they are often bound to oxygen atoms as a result. Therefore, although the carbon atom has lost electrons, it has gained oxygen atoms—hence the term "oxidation." Conversely, compounds that are capable of accepting electrons will also tend to lose oxygen atoms, leaving them in a reduced state—hence the term "reduction." Oxidation-reduction reactions play a huge role in the capture of energy, in the form of sugar during photosynthesis and adenosine triphosphate (ATP) during respiration.

For more information on this concept, be sure to focus on

- Section 8.2, *Energy is extracted from food through a series of oxidation-reduction reactions*
- Figure 8.6, Oxidation and Reduction Reactions Go Together

Enzyme Catalysis

As described in the chapter, all chemical reactions require a certain amount of activation energy to proceed. The activation energy required prevents the reaction from proceeding without regulation. For many reactions, however, the activation energy can be prohibitive. Cells have found a way to overcome this limitation by employing a special class of proteins, called "enzymes," which work to catalyze, or speed up, a chemical reaction. Catalysis is accomplished by lowering the amount of activation energy required to drive the reaction forward. Enzymes can accomplish this in several ways, including (1) binding to the substrates of the reaction, (2) bringing them in close proximity to one other, or (3) placing stress on chemical bonds. The key concept to remember when discussing enzymes is that, although they play a key role in driving chemical reactions, they are not consumed during the reaction. This means a single enzyme molecule can catalyze the same reaction multiple times, remaining unchanged throughout the process. Another concept to keep in mind is that

enzymes are highly specific—they typically drive one, and only one, chemical reaction.

For more information on this concept, be sure to focus on

- Section 8.3, Enzymes
- Figure 8.7, Enzymes as Molecular Matchmakers

TYING IT ALL TOGETHER

Several concepts presented in this chapter build on those presented in previous chapters and will also be revisited and discussed in greater detail in subsequent chapters, including

Chemical Structures Found in Cells

- Chapter 5—Section 5.6, The Chemical Building Blocks of Life

ATP: Energy for the Cell

- Chapter 5—Section 5.10, Nucleotides and Nucleic Acids

Chloroplast and Energy Capture

- Chapter 6—Section 6.4, *Chloroplasts capture energy from sunlight*
- Chapter 9—Section 9.3, Photosynthesis: Capturing Energy from Sunlight

Mitochondria and Energy Production

- Chapter 6—Section 6.4, *Mitochondria power the cell*
- Chapter 9—Section 9.4, Cellular Respiration: Breaking Down Molecules for Energy

Metabolism

- Chapter 26—Section 26.3, Maintaining the Internal Environment: Homeostasis

PRACTICE QUESTIONS

Factual Knowledge

1. Consider the electric blender used to produce a fruit smoothie in the beginning of the chapter. The use of one form of energy (electricity) to drive another process (kinetic energy of blade movement) illustrates
   a. the first law of thermodynamics.
   b. the second law of thermodynamics.
   c. both laws of thermodynamics.

d. neither law of thermodynamics.

e. metabolism.

2. Metabolic reactions that break down complex molecules into smaller compounds, thereby releasing usable energy for the cell, are best described as
   a. biosynthetic.
   b. catabolic.
   c. catalytic.
   d. photosynthetic.
   e. enzymatic.

3. The ultimate source of energy for living systems is
   a. glucose.
   b. oxygen.
   c. sunlight.
   d. carbon dioxide (CO_2).
   e. none of the above

4. Which of the following statements about cellular energy needs is *false*?
   a. Cell disorder will always tend to increase without a continuous input of energy.
   b. Because the laws of thermodynamics dictate that energy cannot be created or destroyed, cells are forced to acquire energy from external forces.
   c. Many cellular reactions have an associated activation energy.
   d. The most usable energy for cells comes directly from the rapid combustion of glucose.
   e. all of the above

5. The contraction of muscle cells is considered a _____ process that is driven by the release of _____ energy stored in a molecule of ATP.
   a. potential; kinetic
   b. kinetic; potential
   c. potential; mechanical
   d. kinetic; glucose
   e. none of the above

6. Oxygen is a powerful oxidizer because of its electron-repelling power. (True or False)

7. The production and breakdown of _____ is often coupled with the metabolic reactions of biosynthesis and catabolism.
   a. aspirin
   b. DNA
   c. ATP
   d. methane
   e. CO_2

8. Which of the following statements about enzymes is true?
   a. Enzymes do not alter the total energy yield for a reaction.

b. Enzymes are proteins whose three-dimensional shape is key to their function.
   c. Enzymes speed up reactions by lowering activation energy.
   d. Enzymes are highly specific for the reactions they catalyze.
   e. all of the above

9. Carbonic anhydrase is an enzyme that speeds the capture of CO_2 in photosynthesis, thus facilitating carbon transfer from plants to animals. (True or False)

10. One strategy that makes metabolic pathways more efficient is
    a. that enzymes for a given pathway are located in the same place within the cell.
    b. that all of the reactions use the same substrate.
    c. that all of the reactions use the same enzyme.
    d. that enzymes for a pathway are separated into different organelles.
    e. none of the above

11. Which of the following statements regarding the relationship between metabolism and life expectancy is *false*?
    a. In general, larger animals with a slower metabolic rate have a longer life expectancy than smaller animals with a higher metabolic rate.
    b. The rate of metabolism is inversely related to overall life span.
    c. The accumulation of free radicals produced during metabolic reactions can increase life span by protecting critical cellular components.
    d. In humans, women typically have a lower metabolic rate and higher life expectancy.
    e. All of the above are false.

12. Anabolism usually _____ energy for the reaction to occur to produce polymers, whereas catabolism _____ energy to produce monomers.
    a. produces; requires
    b. requires; produces
    c. stores; produces
    d. produces; stores
    e. none of the above

13. Match each term with the best description.
    __ active site
    __ catalyst
    __ metabolism
    __ reduction
    __ substrate
    a. capture and use of energy in organisms
    b. converted into product through the action of an enzyme
    c. critical three-dimensional space on an enzyme

d. speeds up the rate of a reaction
e. result of molecule gaining electrons

Conceptual Understanding

1. A metabolic pathway requires ATP for energy and enzymes for catalyzing the production of biological building blocks. (True or False)

2. Which of the following would *not* be considered a type of metabolism?
    a. oxidation of methane molecules
    b. catabolic pathways that break down complex carbohydrates
    c. the capture of light energy for use in making glucose
    d. the manufacture of ATP from the breakdown of glucose
    e. all of the above

3. Your dorm room or apartment will stay organized only if you expend sufficient energy to keep it straightened up. Likewise, cells can remain in a highly ordered state only if they have a supply of energy to use for cellular work. These two examples clearly illustrate
    a. the first law of thermodynamics.
    b. the second law of thermodynamics.
    c. reduction and oxidation.
    d. enzyme specificity.
    e. all of the above

4. A molecule that has been oxidized
    a. has gained an oxygen atom.
    b. has gained one or more hydrogen atoms.
    c. has lost one or more electrons.
    d. a and b
    e. a and c

5. A common way in which cells capture the energy released during the breakdown of large molecules is to add electrons to smaller, specialized molecules that can accept them. This process of electron acceptance is otherwise known as
    a. biosynthesis.
    b. metabolism.
    c. reduction.
    d. catalysis.
    e. a pathway.

6. Figure 8.8 in your textbook graphically illustrates the concept of activation energy for a given chemical reaction. In a chemical reaction, enzymes work to lower the activation energy. Which of the following statements regarding the role of enzymes is incorrect?
    a. Enzymes work by increasing the heat available to a chemical reaction.
    b. Enzymes work by bringing substrates closer together.
    c. Enzymes work by placing strain on the chemical bonds present in substrates.
    d. Enzymes remain unchanged following participation in a chemical reaction.
    e. all of the above

7. Before they can react, many molecules must first be destabilized. This state is typically achieved through
    a. changing the chemical composition of these molecules.
    b. oxidizing the molecules by adding electrons.
    c. changing the reaction from a biosynthetic to a catabolic pathway.
    d. the input of a small amount of activation energy.
    e. none of the above

8. In Figure 8.7a, the object labeled "enzyme" is
    a. a three-dimensional protein.
    b. one component of a larger metabolic pathway.
    c. involved in lowering the activation energy of the reactants to which it binds.
    d. unchanged by its participation in the reaction.
    e. all of the above

9. In the enzyme-catalyzed reaction where A = (Enzyme 1) → B → (Enzyme 2) → C → (Enzyme 3) → D, what will be the effect of inactivating the Enzyme 2?
    a. B, C, and D will all still be produced.
    b. B and C will still be produced, but not D.
    c. B will still be produced, but not C or D.
    d. A will still be produced, but not B, C, or D.
    e. None of the labeled substances will be produced.

10. Enzyme-driven metabolic pathways can be made more efficient by
    a. concentrating enzymes within specific cellular compartments.
    b. grouping enzymes into free-floating, multienzyme complexes.
    c. fixing enzymes into membranes so that they are adjacent to each other.
    d. having the product of one reaction become the reactant of the next.
    e. all of the above

11. Which of the following statements regarding the enzyme carbonic anhydrase is *false*?
    a. Without it, CO_2 would quickly accumulate in the bloodstream.
    b. It uses the compound bicarbonate (HCO_3-) as a substrate.
    c. It helps the body send CO_2 generated in the tissues of the body to the lungs.

    d. It is capable of processing more than 10,000 molecules of CO_2 each second.

    e. all of the above

12. Metabolic rate has been shown to be related to the life span of organisms. Which of the following statements regarding this observation is *false*?

    a. Lower metabolic rates are associated with a longer life span.

    b. Higher metabolic rates are associated with a higher production of toxic by-products.

    c. Metabolic rate can be controlled by restricting the organism's diet.

    d. In general, smaller organisms have slower metabolisms than larger organisms.

    e. Genetics, along with metabolism, plays a role in determining life span.

RELATED ACTIVITIES

- Refer to the chapter's beginning article, "Kick-Start Your Metabolic Engine!" Do some basic research in magazines and online, and make a list of all of the "new fads" having to do with increasing metabolism. Find an article you find compelling and persuasive, and have a nonscientist read it. Discuss with this person the science behind metabolism and the most effective way possible to increase metabolism. How can a higher metabolism affect a human life? List at least three positive and three negative effects.

- This chapter's Biology Matters box, "Enzymes in Action," discusses enzymes that the human body uses and interacts with on a daily basis. Lactose intolerance is very common. What are the side effects if someone who is lactose intolerant consumes lactose? What could be some other enzymes that the body produces that could be depleted over time? Name at least three, then give the effects of this enzyme being depleted. There are some diseases or disorders that cause some enzymes to be inactive. Name one such disease or disorder and the cause and effect of the enzyme inactivity.

- Whenever you cut a potato, an apple, or other fresh fruit or vegetable, before long the exposed tissue will turn brown. This color change is mediated by an enzyme in the plant tissue called "catecholase," which converts a colorless substrate known as "catechol" into a colored product called "benzoquinone" using an oxidation reaction. One may prevent cut fruit from browning by covering it with lemon juice or submerging it in water (H_2O). Suggest a biological explanation that involves enzyme function to explain why this preservation technique works.

ANSWERS AND EXPLANATIONS

Factual Knowledge

1. a. The electricity provided to the blender is used to create the kinetic energy of blade movement, which drives the blending process. This example demonstrates energy conversions governed by the first law of thermodynamics. For more information, see Section 8.1, The Role of Energy in Living Systems and *The laws of thermodynamics apply to living systems.*

2. b. This is the definition of catabolic reactions. For more information, see Section 8.2, *ATP delivers energy to anabolic pathways and is regenerated via catabolic pathways* and *Energy is extracted from food through a series of oxidation-reduction reactions.*

3. c. Although glucose is used by cells to produce ATP, it is ultimately sunlight that provides the energy needed to make glucose through the process of photosynthesis. For more information, see Section 8.1, *The flow of energy connects living things with the environment.*

4. d. It is the slow, stepwise combustion of glucose that makes the harvest of energy so efficient. This process provides the cell with most of its usable energy. For more information, see Section 8.1, *The flow of energy connects living things with the environment.*

5. b. Recall that kinetic energy is the energy of motion or movement, whereas potential energy is the "hidden" energy stored in the chemical bonds of food molecules or ATP. For more information, see Section 8.1, *The laws of thermodynamics apply to living systems.*

6. False. Oxygen is a power oxidizer because of its electron-drawing power. For more information, see Section 8.2, *Energy is extracted from food through a series of oxidation-reduction reactions.*

7. c. ATP is the currency of metabolism. It is frequently made or used in conjunction with other metabolic reactions. For more information, see Section 8.2, *ATP delivers energy to anabolic pathways and is regenerated via catabolic pathways,* and Figure 8.5, Anabolism and Catabolism.

8. e. All of these statements about enzymes are true. For more information, see Section 8.3, Enzymes, *Enzymes remain unaltered and are reused in the course of a reaction,* and *Enzymes increase reaction rates by lowering the activation energy barrier.*

9. False. The action of carbonic anhydrase has nothing to do with photosynthesis. It is an enzyme that catalyzes the conversion of H_2O and CO_2 into HCO_3^- ions in your blood. For more information, see Section 8.3, *Enzymes remain unaltered and are reused in the course of a reaction,* and Figure 8.7b, The Action of Carbonic Anhydrase.

10. a. Concentrating both substrates and enzymes in the same cellular location is one way to make metabolic pathways more efficient. Other ways include attaching enzymes to membranes or combining them into free-floating multienzyme complexes. For more information, see Section 8.4, Metabolic Pathways.

11. c. Recall that free radicals are toxic by-products that are generated during metabolic reactions. They produce detrimental effects on cells by oxidizing critical cellular components such as DNA and lipids, resulting in their damage. For more information, see Section 8.4, Metabolic Pathways, and the Applying What We Learned article, "Metabolic Rates, Health, and Longevity," at the end of the chapter.

12. b. For more information, see Section 8.2, *ATP delivers energy to anabolic pathways and is regenerated via catabolic pathways*, and Figure 8.4, Metabolism.

13. c. active site. For more information, see Section 8.3, *The shape of an enzyme determines its function*.
    d. catalyst. For more information, see Section 8.3, Enzymes.
    a. metabolism. For more information, see Section 8.2, Metabolism.
    e. reduction. For more information, see Section 8.2, *Energy is extracted from food through a series of oxidation-reduction reactions*.
    b. substrate. For more information, see Section 8.3, *Enzymes remain unaltered and are reused in the course of a reaction*.

Conceptual Understanding

1. True. A metabolic pathway must have rich energy, such as ATP, for the catalysis of any molecule with the help of enzymes to form new biological molecules. For more information, see Section 8.2, *ATP delivers energy to anabolic pathways and is regenerated via catabolic pathways*, and Section 8.4, Metabolic Pathways.

2. a. Although the oxidation of methane is a type of catabolic reaction, it is not considered a type of metabolic reaction that would occur naturally inside an organism. For more information, see Section 8.2, *ATP delivers energy to anabolic pathways and is regenerated via catabolic pathways*, and *Energy is extracted from food through a series of oxidation-reduction reactions*.

3. b. The second law of thermodynamics states that the disorder of a system will increase without an input of energy, whether that system be a cell or your room. For more information, see Section 8.1, *The laws of thermodynamics apply to living systems*.

4. e. Recall that the classical description of oxidation refers to a situation in which an atom or molecule gains an oxygen atom. In organic systems, the gain of an oxygen atom by carbon would be accompanied by the loss of one or more hydrogen atoms, and conversely electrons, in the process. Therefore, both answers "a" and "c" apply for an oxidation reaction. For more information, see Section 8.2, *Energy is extracted from food through a series of oxidation-reduction reactions*.

5. c. Reduction is the transfer of energy through the acceptance of electrons by a molecule. It is the opposite of oxidation, in which electrons are released. For more information, see Section 8.2, *Energy is extracted from food through a series of oxidation-reduction reactions*.

6. a. Recall that enzymes work as a biological catalyst inside the cell. Although catalysts act to increase the rate of a chemical reaction, they do not provide any additional energy to drive the reaction forward. For more information, see Figure 8.8, Enzyme Catalysis, and Section 8.3, *Enzymes increase reaction rates by lowering the activation energy barrier*.

7. d. This small amount of added energy is usually sufficient to overcome the activation energy barrier and destabilize the molecule. For more information, see Section 8.2, *Chemical reactions are governed by the laws of thermodynamics*.

8. e. All of these statements about enzymes are true. For more information, see Figure 8.7a, Enzyme Action, and Section 8.3, *The shape of an enzyme determines its function*.

9. c. Because Enzyme 2 catalyzes the conversion of B into C, no C will be made. This means that Enzyme 3 cannot convert C into D, so no D will be made either. However, because Enzyme 1 is still functional, A will be converted into B. For more information, see Section 8.4, Metabolic Pathways.

10. e. All of these are mechanisms used by cells to increase the efficiency of enzymatic pathways in metabolism. For more information, see Section 8.4, Metabolic Pathways, and Figure 8.9, Arrangement of Metabolic Pathways.

11. b. Recall that carbonic anhydrase catalyzes the reaction in which H_2O and CO_2 act as substrates to generate HCO_3^- as a product. Bicarbonate carries the extra carbon to the lungs, where it is converted back to CO_2 and released into the air during the process of exhaling. For more information, see Section 8.3, *Enzymes remain unaltered and are reused in the course of a reaction* and *The shape of an enzyme determines its function*, and Figure 8.7b, Enzyme Action.

12. d. Recall that smaller animals generally have faster metabolic rates and, therefore, shorter life spans than larger animals. For more information, see Section 8.4, Metabolic Pathways, and the Applying What We Learned article, "Metabolic Rates, Health, and Longevity."

CHAPTER 9 | Photosynthesis and Cellular Respiration

GETTING STARTED

Below are a few questions to consider before reading Chapter 9. These questions will help guide your exploration and assist you in identifying some of the key concepts presented in this chapter.

1. What is the relationship between the energy carriers adenosine triphosphate (ATP), nicotinamide adenine dinucleotide (NADH), and nicotinamide adenine dinucleotide phosphate (NADPH)?

2. How are the processes of photosynthesis and cellular respiration interrelated?

3. Why are the leaves of most plants green?

4. How are proton gradients used in cells to generate chemical energy?

5. Under what conditions are the products ethanol and lactic acid produced by organisms?

6. What role does oxygen (O_2) play in cellular respiration?

A GUIDE TO THE READING

The following concepts typically give students the most difficulty when exploring the content in Chapter 9 for the first time. For each concept, one or more references have been identified that may help you gain a better understanding of these potentially problematic areas.

Phosphorylation

The energy carrier molecule ATP is often used to transfer energy to organic compounds within the cell. When one of the phosphate atoms from the ATP molecule is transferred to an organic compound, this compound is said to be "phosphorylated." Phosphorylation of molecules can result in changes in their shape or in activation or deactivation of enzymatic activity. This process plays a key role in regulating the activities of enzymes within the cell. It is important to note that the cell utilizes a process called "oxidative phosphorylation" to regenerate ATP during cellular respiration. In this process, the energy stored in the proton gradient established across the inner mitochondrial membrane is used to drive the activity of the enzyme ATP synthase. ATP synthase is responsible for phosphorylating ADP molecules to generate ATP.

For more information on this concept, be sure to focus on

- Section 9.1, Molecular Energy Carriers
- Section 9.4, *Oxidative phosphorylation uses oxygen to produce ATP in quantity*
- Figure 9.13, Oxidative Phosphorylation

Reactions of Photosynthesis

Photosynthesis, which occurs in the chloroplast, can be broken down into two interrelated processes—light reactions and the Calvin cycle. The light reactions of photosynthesis work to capture the energy of sunlight inside the chloroplast. As we saw in the chapter, the pigment chlorophyll is green in color; this is because it reflects green light, causing our eyes to see green when we look at a leaf. Consequently, the pigment is efficient at absorbing both red-orange and blue-violet light. When light is absorbed, electrons present in the chlorophyll molecule become "energized" and are then passed on to an electron transport chain (ETC) that is arranged in one of two types of photosystems (I and II). Photosystem I accepts high-energy electrons and produces NADPH, a compound used in the cell to reduce compounds.

Photosystem II accepts electrons and participates in photolysis, or the splitting of water (H_2O). In either case, the transport of electrons in the ETC helps generate a proton gradient across the thylakoid membrane in the chloroplast, which can then be used to generate ATP in the cell. The ATP and NADPH produced by the light reactions of photosynthesis are used by the cell to drive the Calvin cycle of photosynthesis. In the Calvin cycle, carbon dioxide (CO_2) from the air is combined with H_2O to produce glucose molecules and O_2. This process requires energy obtained from both ATP and hydrogen ions donated by NADPH. In contrast to the light reactions, the Calvin cycle of photosynthesis does not require light to proceed.

For more information on this concept, be sure to focus on

- Section 9.3, *The light reactions generate energy carriers*
- Section 9.3, *The Calvin cycle reactions manufacture sugars*
- Figure 9.7, Light Harvesting and Electron Flow
- Figure 9.8, The Light Reactions
- Figure 9.9, The Calvin Cycle Converts Inorganic Carbon into Sugar

The Krebs Cycle

During cellular respiration, the process of glycolysis is followed by a stepwise series of reactions that donate high-energy electrons to the compound NAD^+, generating NADH. As we saw in the chapter, the process of glycolysis results in the generation of two molecules of pyruvate for every glucose molecule used. The three-carbon pyruvate releases a single CO_2 molecule and joins to a compound called "coenzyme A." The resulting molecule is called "acetyl CoA." Acetyl CoA then enters the Krebs cycle, which is a series of eight oxidation reactions that occur within the mitochondrial matrix. As the Krebs cycle proceeds, carbon atoms are released as CO_2 (and then exhaled) and high-energy electrons are stored in NADH. The NADH molecules produced from the Krebs cycle are then used to drive the final stage of respiration, oxidative phosphorylation, which produces ATP. It is important to note that very little ATP is produced during the Krebs cycle. The primary product is, in fact, the high-energy NADH molecules.

For more information on this concept, be sure to focus on

- Section 9.4, *The Krebs cycle releases carbon dioxide and generates energy carriers*
- Figure 9.12, The Krebs Cycle

TYING IT ALL TOGETHER

Several concepts presented in this chapter build on those presented in previous chapters and may also be revisited and discussed in greater detail in subsequent chapters, including

Mitochondrial and Chloroplast Structure

- Chapter 6—Section 6.4, Internal Compartments of Eukaryotic Cells

Diffusion

- Chapter 7—Section 7.1, The Plasma Membrane as Gate and Gatekeeper

Oxidation-Reduction Reactions

- Chapter 8—Section 8.2, *Energy is extracted from food through a series of oxidation-reduction reactions*

Metabolism

- Chapter 8—Section 8.2, Metabolism
- Chapter 26.3—Section 26.3, Maintaining the Internal Environment: Homeostasis

Digestion and the Breakdown of Food

- Chapter 27—Section 27.1, Nutrients That Animals Need

PRACTICE QUESTIONS

Factual Knowledge

1. Which of the following statements about energy metabolism is *false*?
   a. The energy that powers living systems ultimately comes from the sun.
   b. All animals in some way rely on plants for their energy.
   c. Plants provide the H_2O and CO_2 that animals need to carry out respiration.
   d. Most eukaryotic organisms carry out respiration in the presence of O_2.
   e. The main objective of respiration is to break down glucose and make ATP.

2. Which of the following statements regarding photosynthesis and cellular respiration is *false*?
   a. Photosynthesis is an anabolic process, whereas cellular respiration is a catabolic process.
   b. Oxygen is required in cellular respiration to assist in the oxidation of glucose.
   c. The process of photosynthesis does not require the input of energy.
   d. Photosynthesis fixes CO_2 during the Calvin cycle, whereas cellular respiration releases CO_2 during the Krebs cycle.
   e. none of the above

3. During which cellular process is ATP released?
   a. Calvin cycle of photosynthesis
   b. Krebs cycle of cellular respiration
   c. glycolysis
   d. a and b
   e. b and c

4. The stroma is located in the mitochondrion and is the place where the light reactions of photosynthesis take place. (True or False)

5. Which of the following are logically associated with chloroplasts?
   a. plant cells
   b. Calvin cycle
   c. chlorophyll
   d. thylakoid membranes
   e. all of the above

6. Antenna complexes, ETCs, and carbon fixation are all found in
   a. animal cells.
   b. bacterial cells.
   c. plant cells.
   d. mitochondria.
   e. all of the above

7. The manufacture of ATP in both photosynthesis and cellular respiration is made possible by
   a. the existence of a proton gradient across specific membranes.
   b. the action of ATP synthase.
   c. energy from the movement of electrons.
   d. the action of electron-accepting proteins.
   e. all of the above

8. Rubisco, glyceraldehyde 3-phosphate, and NADPH all play a role in
   a. the light reactions of photosynthesis.
   b. the Calvin cycle of photosynthesis.
   c. the breakdown of glucose into CO_2.
   d. cellular respiration when O_2 is present.
   e. alcohol fermentation.

9. Six molecules of CO_2 must be captured during the Calvin cycle of photosynthesis for every molecule of glucose that is manufactured. (True or False)

10. Which of the following represents the correct ordering of events that occur during the catabolism of glucose in the absence of O_2?
    a. glycolysis; Krebs cycle; oxidative phosphorylation
    b. glycolysis; oxidative phosphorylation; Krebs cycle
    c. oxidative phosphorylation; Krebs cycle; glycolysis
    d. glycolysis; fermentation
    e. Krebs cycle; fermentation

11. Glycolysis takes place in the _____ and produces _____, which in the presence of O_2 then enters the _____.
    a. cytosol; glucose; mitochondrion to complete fermentation
    b. cytosol; pyruvate; mitochondrion to complete fermentation
    c. cytosol; pyruvate; mitochondrion to complete cellular respiration
    d. mitochondrion; pyruvate; chloroplast to complete photosynthesis
    e. chloroplast; glucose; cytosol to complete cellular respiration

12. During cellular respiration, _____ is the step that produces the greatest number of ATP molecules.
    a. fermentation
    b. glycolysis
    c. Krebs cycle
    d. oxidative phosphorylation
    e. carbon fixation

13. Both NAD^+ and $NADP^+$
    a. are reduced.
    b. have a full complement of electrons.
    c. are oxidized.
    d. are used during carbon fixation in photosynthesis.
    e. none of the above

14. Match each term with the best description.
    __ aerobic
    __ catabolism
    __ energy carrier
    __ glycolysis
    __ thylakoids
    a. in the presence of O_2
    b. takes place in the cytosol
    c. cellular respiration, for example
    d. embedded within the stroma
    e. NADH

Conceptual Understanding

1. Refer to Figure 9.1 in your textbook. In terms of energy, what is the overall purpose of the two sets of metabolic processes shown?
   a. to make glucose out of sunlight energy
   b. to convert sunlight energy into ATP
   c. to fix carbon into biological molecules
   d. to make O_2 so that organisms can respire
   e. to recycle CO_2 between plants and animals

2. Which of the following ultimately provides the O_2 used during the process of oxidative phosphorylation?
   a. fermentation
   b. light reactions of photosynthesis

c. Calvin cycle of photosynthesis
d. Krebs cycle
e. carbon fixation

3. Imagine that you have been shrunk to microscopic size and are sitting on top of a thylakoid disk. Which of the following are contained within the structure upon which you sit?
   a. photosystems
   b. antenna complexes
   c. ETCs
   d. phospholipids
   e. all of the above

4. Which of the following serves as both reactant in photosynthesis and product in cellular respiration?
   a. O_2
   b. CO_2
   c. H_2O
   d. a and b
   e. b and c

5. The protons that make up the proton gradient used during the light reactions of photosynthesis come from
   a. glucose.
   b. ATP.
   c. H_2O.
   d. NADPH.
   e. rubisco.

6. The electrons that are released by the splitting of H_2O molecules during photosynthesis ultimately end up in
   a. glucose.
   b. ATP.
   c. H_2O.
   d. NADPH.
   e. rubisco.

7. What process in cellular respiration is essentially the reverse of carbon fixation in photosynthesis?
   a. glycolysis
   b. Krebs cycle
   c. oxidative phosphorylation
   d. alcohol fermentation
   e. lactic acid fermentation

8. To carry out glycolysis, a cell must have glucose and mitochondria. (True or False)

9. Acetyl coenzyme A (CoA), CO_2, oxaloacetate, and NADH are all
   a. components or products of the Krebs cycle.
   b. part of the Calvin cycle of photosynthesis.
   c. part of the reactions of lactic acid fermentation.
   d. elements of oxidative phosphorylation.
   e. all of the above

10. The muscle cells of white lab rats have 50 percent more mitochondria than do those of burrowing naked mole rats. (True or False)

11. If you consider the combined processes of photosynthesis and cellular respiration, the electrons found in H_2O at the beginning of the light reactions end up attaching to _____ during respiration.
    a. O_2 to make new H_2O
    b. NADPH to make new glucose
    c. pyruvate to make ethanol
    d. electron transport carriers to make O_2
    e. none of the above

12. Consider a hypothetical eukaryotic cell that has only a single glucose molecule available for respiration. If this cell can perform only fermentation instead of cellular respiration, it
    a. will produce fewer CO_2 molecules as a result.
    b. will generate fewer ATP molecules as a result.
    c. probably lacks O_2.
    d. will have mitochondria present.
    e. all of the above

13. Oxidative phosphorylation is to respiration as _____ is to photosynthesis.
    a. carbon fixation
    b. the splitting of H_2O
    c. the ETC
    d. light capture by chlorophyll
    e. the reduction of NADPH

RELATED ACTIVITIES

- The article "Waiting to Exhale," presented at the end of the chapter, suggests that those who partake in sports and competitions that require less O_2 can actually use less ATP and decrease their metabolic activity. Determine at least five different ways you could decrease your metabolic activity in hopes of decreasing O_2 necessity. How could performing these activities for prolonged periods affect the human body over time?

- Each of your body's cells uses up its store of ATP every minute or two. It should therefore be apparent to you just how critical the role of oxidative phosphorylation is to your body, because that is where the vast majority of ATP comes from. With this in mind, search the Internet for information on the fast-acting poison called "cyanide." Compose a one-page essay discussing the relationship between mitochondria, oxidative phosphorylation, and the action of cyanide.

- Using the steps of the scientific method, propose a hypothesis and a method of carrying out an experiment to determine the rate of photosynthesis and cellular respiration. What would the experiment setup be? How could these rates be safely determined and then analyzed?

ANSWERS AND EXPLANATIONS

Factual Knowledge

1. c. Cellular respiration does not use H_2O or CO_2 as reactants; these are actually two of the products of respiration. Plants provide the glucose and O_2 that make respiration possible. For more information, see the Chapter 9 opener, "Energy Is Necessary."

2. c. Recall that the process of photosythesis is driven by the energy of sunlight. This energy is used to assist in the process of fixing CO_2 to produce glucose. For more information, see Section 9.2, *Light powers the manufacture of carbohydrates during photosynthesis* and *Energy from sugars is used to make ATP during cellular respiration.*

3. e. While the majority of ATP produced during cellular respiration is generated from the process of oxidative phosphorylation, some ATP is released during the Krebs cycle as well as in the process of glycolysis. For more information, see Section 9.2, *Energy from sugars is used to make ATP during cellular respiration.*

4. False. This statement is wrong on both counts. The stroma is part of the chloroplast, and it is where the Calvin cycle of photosynthesis takes place. For more information, see Section 9.3, *The light reactions generate energy carriers*, and Figure 9.6, Chloroplast Structure.

5. e. Chloroplasts are found in plant cells and contain both thylakoid membranes and chlorophyll. The Calvin cycle takes place within the stroma of the chloroplast. For more information, see Section 9.3, *Chloroplasts are photosynthetic organelles* and *The Calvin cycle reactions manufacture sugars*, and Figure 9.6, Chloroplast Structure.

6. c. These are all participants or processes associated with photosynthesis, which occurs in plant cells. For more information, see Section 9.3, *The light reactions generate energy carriers.*

7. e. As electrons move through the electron acceptor molecules present on the inner membranes of both chloroplasts and mitochondria, the energy associated with the electrons is used to generate a proton gradient. Protons are then passed through an ATP synthase to provide the energy needed to manufacture ATP. For more information, see Figure 9.8, The Light Reactions, and Figure 9.13, Oxidative Phosphorylation.

8. b. All three are players in the Calvin cycle of photosynthesis, where carbon fixation occurs. Rubisco is an enzyme that catalyzes the capture of CO_2, glyceraldehyde 3-phosphate is one of the intermediates in the chemical pathway that leads to glucose, and NADPH is an electron carrier made during the light reactions that is used to power the Calvin cycle. For more information, see Section 9.3, *The Calvin cycle reactions manufacture sugars*, and Figure 9.9, The Calvin Cycle.

9. True. There are six carbons in a glucose molecule, so six molecules of CO_2 must be captured for each glucose molecule made. For more information, see Section 9.3, *The Calvin cycle reactions manufacture sugars.*

10. d. In the absence of O_2, all that can occur is glycolysis and fermentation. Oxygen must be present for the Krebs cycle and oxidative phosphorylation to take place. For more information, see Section 9.4, *Fermentation facilitates ATP production through glycolysis when oxygen is absent.*

11. c. For a summary of these possibilities, see Figure 9.10, Glycolysis, and Figure 9.11, Fermentation.

12. d. Over 80 percent of the ATP produced during cellular respiration is the result of oxidative phosphorylation. For more information, see Section 9.4, *Oxidative phosphorylation uses oxygen to produce ATP in quantity.*

13. c. The positive sign indicates a charge that results from the loss of electrons, which is what makes these molecules oxidized. For more information, see Section 9.1, Molecular Energy Carriers. Also refer to Chapter 5.

14. a. aerobic. For more information, see Section 9.4, *Cellular respiration in the mitochondrion furnishes much of the ATP needed by most eukaryotes.*
    c. catabolism. For more information, see Section 9.1, Molecular Energy Carriers.
    e. energy carrier. For more information, see Section 9.1, Molecular Energy Carriers.
    b. glycolysis. For more information, see Section 9.4, *Glycolysis is the first stage in the cellular breakdown of sugars.*
    d. thylakoids. For more information, see Section 9.3, *Chloroplasts are photosynthetic organelles.*

Conceptual Understanding

1. b. All of these things happen as a result of some aspect of energy metabolism. However, the overall purpose of the combined processes of photosynthesis and cellular respiration is to take an abundant, but unusable, form of energy (sunlight) and convert it to a form of energy that is usable by the cell (ATP). For more information, see Section 9.1, Molecular Energy Carriers.

2. b. Oxygen is made when H_2O molecules are split to generate electrons during the light reactions of photosynthesis. For more information, see Section 9.4, *Oxidative phosphorylation uses oxygen to produce ATP in quantity*, and Section 9.3, *The light reactions generate energy carriers*.

3. e. The thylakoid disk is part of the inner membrane of the chloroplast and is involved with the light reactions of photosynthesis. Therefore, all of these structures would be a part of the disk. For more information, see Section 9.3, *The light reactions generate energy carriers* and *Chloroplasts are photosynthetic organelles*.

4. e. Both H_2O and CO_2 are used during photosynthesis to help make glucose, and are produced during the breakdown of glucose that occurs during cellular respiration. For more information, see Section 9.2, *Light powers the manufacture of carbohydrates during photosynthesis* and *Energy from sugars is used to make ATP during cellular respiration*.

5. c. The splitting of H_2O molecules in the light reactions gives rise to the protons needed for this gradient. For more information, see Section 9.3, *The light reactions generate energy carriers*.

6. d. The reason for splitting H_2O molecules is to provide electrons to the photosystems used during the light reactions. These photosystems need new electrons because some of them are used to reduce $NADP_+$ into NADPH. For more information, see Section 9.3, *The light reactions generate energy carriers*.

7. b. Carbon fixation converts CO_2 into glucose, and the Krebs cycle essentially reverses this process. For more information, see Section 9.4, *The Krebs cycle releases carbon dioxide and generates energy carriers*, and Section 9.3, *The Calvin cycle reactions manufacture sugars*.

8. False. Glycolysis occurs in the cytosol of the cell. For more information, see Figure 9.10.

9. a. These are all components of the Krebs cycle. For more information, see Figure 9.12.

10. False. Just the reverse is true. Naked mole rats live in oxygen-poor underground tunnels. They spend much of their time digging through hard soils and therefore have a great demand for ATP. To generate the necessary ATP in a low-O_2 environment, the muscle cells of the mole rat have adapted by accumulating more mitochondria than those of lab rats. For more information, see Section 9.4, *Cellular respiration in the mitochondrion furnishes much of the ATP needed by most eukaryotes*.

11. a. It is interesting that the electrons both begin their journey through metabolism in H_2O and end that journey in the water that is made when O_2 serves as the final electron acceptor in oxidative phosphorylation, before combining with protons to form new "metabolic" water. For more information, see Section 9.3, *The light reactions generate energy carriers*; Section 9.4, *Oxidative phosphorylation uses oxygen to produce ATP in quantity*; Figure 9.8, The Light Reactions; and Figure 9.13, Oxidative Phosphorylation.

12. e. The lack of O_2 is what directs the products of glycolysis into fermentation instead of into the mitochondrion for use in aerobic respiration. This will result in reduced CO_2 and ATP production. For more information, see Section 9.4, *Glycolysis is the first stage in the cellular breakdown of sugars* and 9.4, *Fermentation facilitates ATP production through glycolysis when oxygen is absent*.

13. c. Both processes use electron acceptor molecules to carry out a series of oxidation-reduction reactions that generate energy for use in the production of ATP. For more information, see Section 9.3, *The light reactions generate energy carriers*, and Section 9.4, *Oxidative phosphorylation uses oxygen to produce ATP in quantity*.

CHAPTER 10 | Cell Division

Below are a few questions to consider before reading Chapter 10. These questions will help guide your exploration and assist you in identifying some of the key concepts presented in this chapter.

1. What are stem cells, and how are they different from somatic cells found in your body?

2. What are the differences between the G_1 and G_2 phases of the cell cycle?

3. What role do the cytoskeletal structures called "microtubules" play in cell division?

4. What is a cell plate, and what role does it play in the process of plant cell division?

5. What is the primary difference between haploid and diploid cells?

6. How many copies of each chromosome are present in cells that have undergone meiosis I? How about in meiosis II?

7. How many different types of differentiated cells are there in the human body?

A GUIDE TO THE READING

The following concepts typically give students the most difficulty when exploring the content in Chapter 10 for the first time. For each concept, one or more references have been identified that may help you gain a better understanding of these potentially problematic areas.

Homologous Chromosomes

As we saw in the chapter, homologous chromosomes are pairs of chromosomes containing the same set of genes. One homolog of the pair is inherited from the organism's mother, and the other homolog is inherited from the father. Humans possess 23 pairs of homologous chromosomes, for a grand total of 46 individual chromosomes. The process of cell division must utilize a mechanism that ensures that each daughter cell obtains a complete set of 23 pairs of homologous chromosomes. To accomplish this, each individual chromosome must be copied. This occurs during the S phase of the cell cycle. After a chromosome is copied, the two duplicate strands of DNA (which are physically joined together by a structure located in the center, called a "centromere") are referred to as "chromatids." With each of the 46 chromosomes having an attached duplicate, the cell can then begin the process of cell division. By aligning the chromatids in the center of the cell during metaphase, the cell ensures that each side of itself will obtain one copy of each of the 46 chromosomes following cell division.

For more information on this concept, be sure to focus on

- Section 10.3, *Most human cells have two copies of each type of chromosome*
- Figure 10.7, The Packing of DNA into a Chromosome
- Figure 10.9, Human Karyotype
- Figure 10.10, Stages of Mitosis and Cytokinesis

Meiosis

The key to understanding meiosis is knowing the overall goal of the process: to reduce the number of chromosomes in the daughter cells (gametes). Gametes in sexually reproducing species are haploid; that is, they contain only one copy of each chromosome. This process of producing haploid cells requires two cell divisions occurring in sequence, called "meiosis I" and "meiosis II." The key step in the process is the pairing of homologous chromosomes, forming bivalents, during meiosis I. This pairing process does not occur during mitosis. When these bivalents are split during anaphase I, one complete chromosome (consisting of two identical chromatids) is segregated into each of the two daughter cells. The result is that each daughter cell formed after meiosis I has two copies of only one complete set of chromosomes and therefore would be considered haploid. The process of meiosis II then serves to separate the chromatids into one of the two resulting daughter cells. In the end, the process of meiosis produces four gametes, each with only one copy of each of the chromosomes.

For more information on this concept, be sure to focus on

- Section 10.5, Meiosis: Halving the Chromosome Set to Make Gametes
- Figure 10.13, Sexual Reproduction Requires a Reduction in Chromosome Number
- Figure 10.14, The Main Events in Meiosis

Genetic Variation

The complexity of life-forms found on earth can be thought of in terms of the genetic differences that exist among species. However, even within a given species, such as humans, it is clear that tremendous genetic variation exists. This is why no two people, with the exception of identical twins, look exactly the same. Organisms that reproduced sexually have many mechanisms through which they maintain this genetic variability. Chapter 10 describes the process of meiosis, which serves to produce haploid gametes with a single copy of each chromosome. By randomly aligning bivalents in the center of the cell during metaphase I of meiosis, each daughter cell will receive a unique combination of chromosomes. In the case of humans, with 23 unique chromosomes, the total number of possible chromosome combinations that can be produced through meiosis is 2^{23} or 8,388,608. The chapter goes on to describe the process of crossing-over that occurs during prophase I and metaphase I of meiosis, during which bivalents exchange pieces of DNA. This process further increases the genetic variability produced through the process of meiosis and sexual reproduction.

For more information on this concept, be sure to focus on

- Section 10.5, *Meiosis and fertilization contribute to genetic variation in a population*
- Figure 10.15, Crossing-Over Produces Recombinant Chromosomes

TYING IT ALL TOGETHER

Several concepts presented in this chapter build on those presented in previous chapters and may also be revisited and discussed in greater detail in subsequent chapters, including

Cell Communication

- Chapter 7—Section 7.6, Cell Signaling

Chromosomes and Genetics

- Chapter 13—Section 13.1, The Role of Chromosomes in Inheritance

Crossing-Over

- Chapter 13—Section 13.3, Genetic Linkage and Crossing-Over

Human Reproduction

- Chapter 33—Reproduction and Development

PRACTICE QUESTIONS

Factual Knowledge

1. The process through which a newly formed daughter cell becomes different from the original parent cell is
   a. cell differentiation.
   b. binary fission.
   c. cell cycle.
   d. meiosis.
   e. interphase.

2. Certain types of cell growth and repair can occur only with the involvement of a unique class of cells that can divide and remain unspecialized for the entire lifetime of an organism. These cells are known as
   a. gametes.
   b. muscle cells.
   c. skin cells.
   d. stem cells.
   e. cancer cells.

3. Before a cell can enter mitosis, it must first pass through
   a. interphase.
   b. G_1 phase.
   c. G_2 phase.
   d. S phase.
   e. all of the above

4. Which of the following statements about the G₀ phase is true?
   a. Cells in this phase do not have cell cycle regulatory proteins.
   b. It is a phase that prepares the cell for mitosis.
   c. It is a phase that prepares the cell for meiosis.
   d. It is the phase in which DNA is replicated in preparation for cell division.
   e. all of the above

5. _____ is the process whereby an equal amount of parental genetic material is distributed between daughter nuclei destined for separate daughter cells.
   a. DNA duplication
   b. DNA segregation
   c. Meiosis
   d. Late prophase
   e. Cytokinesis

6. A physician suspects that a chromosomal abnormality is the cause of a patient's disease. He orders a karyotype for this patient, which will allow the doctor to
   a. see the genes located on the patient's chromosomes.
   b. determine which chromosomes (maternal or paternal) are present in the patient's cells.
   c. see the number of mutations the patient has.
   d. count the total number of chromosomes present in the patient's cells.
   e. all of the above

7. In which of the following stages of mitosis would you expect to see chromatids become daughter chromosomes?
   a. prophase
   b. prometaphase
   c. metaphase
   d. anaphase
   e. telophase

8. The key feature of metaphase in mitosis is that
   a. the nuclear membrane begins to break down.
   b. new DNA is synthesized.
   c. chromosomes are no longer in a duplicated form.
   d. individual chromosomes line up along the equator of the cell.
   e. all of the above

9. Cytokinesis, the physical act of cell separation, occurs by the same mechanism in both plants and animals. (True or False)

10. The most important characteristic that distinguishes meiosis from mitosis is
    a. the total number of chromosomes present in daughter cells.

    b. the number of times each stage of the division process occurs.
    c. whether the chromosomes are duplicated at the end of cell division.
    d. the presence of metaphase in mitosis, but not in meiosis.
    e. all of the above

11. The reduction division of meiosis specifically occurs at the end of meiosis II. (True or False)

12. The second division of meiosis is essentially
    a. the repeat of the first division of meiosis.
    b. a way to turn haploid cells into diploid cells.
    c. mitosis that begins with two haploid cells and ends with four haploid cells.
    d. mitosis that begins with two diploid cells and ends with four diploid cells.
    e. an unnecessary holdover from the early evolution of cells.

13. Which of the following would *not* be found in cells undergoing anaphase II of meiosis?
    a. daughter chromosomes
    b. homologous pairs of chromosomes
    c. DNA
    d. a haploid number of chromosomes
    e. centromeres

14. Match each term with the best description.
    __ sister chromatids
    __ G₀ phase
    __ haploid
    __ prophase
    __ telophase
    a. the product of DNA replication during S phase
    b. the stage of division in which cytokinesis occurs
    c. the resting stage for the cell in which no cell division occurs
    d. immediately follows the end of interphase
    e. no homologous chromosomes present

Conceptual Understanding

1. The aftereffects of a sunburn, as depicted in Figure 10.1 in your textbook, show how skin cells are replaced as they are lost from the surface of the skin. This replacement process represents an example of
   a. mitosis.
   b. meiosis I.
   c. meiosis II.
   d. meiosis I and II.
   e. none of the above

2. In Figure 10.5, what is the difference between the cell
   shown at the beginning of G$_2$ phase and the one
   shown entering mitosis and cytokinesis?
   a. cell size only
   b. cell size and chromosome number
   c. cell size and whether chromosomes have been
      duplicated
   d. chromosome number only
   e. chromosome number and whether chromosomes
      have been duplicated

3 If you compare the chromosomes of a cell as it enters
   interphase with the same chromosomes just before the
   cell begins mitosis, a key difference would be
   a. the total number of chromosomes that are present.
   b. whether the chromosomes are duplicated.
   c. whether homologous pairs of chromosomes are
      present.
   d. the number of different kinds of chromosomes
      present.
   e. all of the above

4. A cell whose haploid number is 12 will have
   _____ chromosomes present at prometaphase of
   mitosis.
   a. 48
   b. 24
   c. 12
   d. 6
   e. cannot determine from the information given

5. In humans, mitosis
   a. produces an exact genetic copy of the parent cell.
   b. results in diploid daughter cells.
   c. does not produce sperm or eggs.
   d. makes daughter cells whose chromosomes are
      unduplicated.
   e. all of the above

6. A normal diploid cell from a newly identified
   mammal has a total of 50 chromosomes. Which of the
   following statements about this cell is true?
   a. If it undergoes meiosis, the daughter cells will have
      50 chromosomes.
   b. The cell contains 24 homologous pairs of
      chromosomes plus the sex chromosomes.
   c. In metaphase of mitosis, this cell will have 50
      chromatids per cell.
   d. In metaphase II of meiosis, there will still be 50
      chromosomes per cell.
   e. all of the above

7. How many pairs of homologous chromosomes would
   be present in a cell undergoing metaphase II?
   a. four
   b. two

c. one
d. none
e. cannot determine from the information given

8. Consider a cell that begins meiosis with 30 total
   chromosomes. How many chromosomes will be
   present in each resulting daughter cell by the time the
   division process reaches metaphase II?
   a. 0
   b. 15
   c. 20
   d. 30
   e. 60

9. If you randomly removed a somatic cell from the
   human body, odds are that it would
   a. have been produced by meiosis.
   b. be haploid.
   c. have 46 chromosomes.
   d. have 23 chromosomes.
   e. be in the part of the cell cycle called "mitosis and
      cytokinesis."

10. The main difference between plant and animal cell
    division is that
    a. cytokinesis takes place in different ways.
    b. meiosis occurs only in animals.
    c. mitosis occurs only in plants.
    d. plants never produce diploid cells.
    e. metaphase does not occur in plant cell division.

11. In which of the following stages would a human cell
    not be diploid?
    a. prophase of mitosis
    b. telophase of mitosis
    c. metaphase of mitosis
    d. prophase of meiosis I
    e. metaphase of meiosis II

12. You are asked to look at a microscope slide containing
    dividing human cells. Focusing on one of these cells,
    you note that it has 23 unduplicated chromosomes.
    This cell is most likely in which of the following
    stages of cell division?
    a. prometaphase of mitosis
    b. telophase of meiosis II
    c. anaphase of mitosis
    d. metaphase of meiosis I
    e. prophase of meiosis II

13. Which of the following contributes to the overall
    genetic diversity present in an organism?
    a. an independent assortment of chromosomes
    b. genetic recombination
    c. mutation
    d. random fertilization of egg cells by sperm cells
    e. all of the above

RELATED ACTIVITIES

- "Olympic-Class Algal Bloom" at the beginning of the chapter describes how an extraordinarily large algae bloom affected the Olympics. Describe what an algae bloom is, what causes its occurrence, and what can be at least five harmful effects it can have on the body of water and its contents. Why are algae blooms more common in recent times? What are two ways algae bloom occurrences can be decreased?

- Mitosis produces genetically identical cells, whereas meiosis leads to the production of genetically variable offspring. Many organisms use some form of sexual reproduction based on the products of meiosis. Propose reasons why having genetic variability in offspring might be advantageous over producing offspring that are genetic replicas of the parent.

- The chapter's Biology in the News box, "Puzzle Solved: How a Fatherless Lizard Species Maintains Its Genetic Diversity," discusses a lizard that has been reproduced without a father. The lizard is a replicate of the original lizard. What are some pros and cons in having a replication of an organism? How can this affect diversity for the organism and its environment? (Think of predators, changing environment, and the diversity of other organisms.)

ANSWERS AND EXPLANATIONS

Factual Knowledge

1. a. Recall that cell differentiation is an important part of the developmental process through which specialized cell types are produced. As described in the chapter, the human body has about 220 types of specialized, differentiated cell types. For more information, see Section 10.1, *Cell divisions grow, maintain, and reproduce the human body.*

2. d. Refer to Figure 10.1 to see how stem cells are used to replenish skin cells as they are lost. For more information, see Section 10.1, *Why Cells Divide.*

3. e. The entirety of interphase (G1, G2, and S) consists of events that are necessary precursors to mitosis. For more information, see Section 10.2, *DNA is replicated in S phase*, and Figure 10.5, The Cell Cycle.

4. a. G_0 is a stasis in which cells are temporarily or permanently prevented from dividing. When stuck in G_0, cells cannot complete the growth and synthesis portions of interphase and thus are unable to divide. Most cells of the body are at least temporarily stuck in G_0 once adulthood is reached. For more information, see Section 10.2, *Most cell types in the adult body do not divide.*

5. b. "Segregation" means separation. DNA segregation is the equal separation of chromosomal material into daughter cells. This stage occurs during mitosis. For more information, see Section 10.4, Mitosis and Cytokinesis: From One Cell to Two Identical Cells.

6. d. Recall that a karyotype is a pictorial representation of the chromosomes taken from a cell about to undergo division. Refer to Figure 10.9 to see a human karyotype. For more information, see Section 10.3, *The karyotype describes all the chromosomes in a nucleus.*

7. d. Sister chromatids separate and become daughter chromosomes during anaphase of mitosis. Refer to Figure 10.10 to see a diagram of this process. For more information, see Section 10.4, *Chromatids separate during anaphase.*

8. d. See Figure 10.10; the single-file arrangement of chromosomes along the metaphase plate characterizes metaphase. For more information, see Section 10.4, *Chromosomes line up in the middle of the cell during metaphase.*

9. False. Because plant cells have a cell wall, cytokinesis cannot occur through the simple inward pinching of membranes, as it does in animal cells. In plants, a new partition called a "cell plate" wall is first laid down between what will become the daughter cells, and then a membrane forms to complete the separation process. Refer to Figure 10.11 and Figure 10.12 to see a diagram of this process. For more information, see Section 10.4, *The cytoplasm is divided during cytokinesis.*

10. a. Recall that the overall purpose of mitosis is to make genetically identical daughter cells (diploid parent cell to diploid daughter cells), whereas the purpose of meiosis is to reduce the number of chromosomes by half (diploid parent cell to haploid daughter cells). For more information, see Section 10.1, *Eukaryotes use mitosis to generate identical daughter cells* and *Meiosis is necessary for sexual reproduction*, and Table 10.1, Biological Relevance of Cell Division.

11. False. The reduction division that turns a diploid cell into haploid daughter cells occurs in meiosis I. For more information, see Figure 10.14, The Main Events in Meiosis, and Section 10.5, *Meiosis I reduces the chromosome number.*

12. c. The purpose of meiosis II is to separate the attached duplicated sister chromatids present in the haploid cells following meiosis I. This is essentially what mitosis does, only with diploid cells instead of haploid. For more information, see Section 10.5, *Meiosis II segregates sister chromatides into separate daughter cells*, and Figure 10.14, The Main Events in Meiosis.

13. b. Recall that homologous pairs of chromosomes are separated during meiosis I and therefore would not be

present at anaphase II. For more information, see Section 10.5, *Meiosis II segregates sister chromatides into separate daughter cells.*

14. a. sister chromatids. For more information, see Section 10.2, *DNA is replicated in S phase.*

c. G$_0$ phase. For more information, see Section 10.2, *Most cell types in the adult body do not divide.*

e. haploid. For more information, see Section 10.1, *Meiosis is necessary for sexual reproduction.*

d. prophase. For more information, see Section 10.4, *Chromosomes are compacted during early prophase.*

b. telophase. For more information, see Section 10.4, *New nuclei form during telophase.*

Conceptual Understanding

1. a. This is an example of producing new body cells, which is accomplished through mitosis. For more information, see Section 10.1, *Eukaryotes use mitosis to generate identical daughter cells.*

2. a. No changes in the amount of DNA or chromosome number occur after S phase. The only thing that changes is cell size. For more information, see Section 10.2, *DNA is replicated in S phase.*

3. b. The primary purpose of S phase, which is a major portion of interphase, is to duplicate chromosomes in preparation for cell division. For more information, see Section 10.2, *DNA is replicated in S phase.*

4. b. Cells undergoing mitosis are diploid. If the haploid number of a cell is 12, then the diploid number is 24. For more information, see Section 10.4, *Chromosomes are attached to the spindle in late prophase.*

5. e. All of these statements are true of mitosis in humans and in any diploid organism. For more information, see Section 10.4, Mitosis and Cytokinesis: From One Cell to Two Identical Cells.

6. b. A diploid cell with 50 chromosomes has 25 pairs, one of which, for most multicellular organisms, consists of the sex chromosomes. During mitosis the cell will remain diploid, so in metaphase there will be 100 chromatids. Meiosis in this cell will produce haploid daughter cells with only 25 chromosomes. For more information, see Section 10.5, *Gametes contain half the chromosomes found in somatic cells.*

7. d. The first division of meiosis (reduction division) yields haploid cells, so throughout all of meiosis II there are no homologous pairs of chromosomes present in the cells. For more information, see Section 10.5, *Meiosis II segregates sister chromatids into separate daughter cells*, and Figure 10.14, The Main Events in Meiosis.

8. b. By the time the cells reach metaphase II, they have already completed meiosis I (reduction division) and are therefore haploid. Each cell will have 15 chromosomes. For more information, see Section 10.5, *Gametes contain half the chromosomes found in somatic cells* and *Meiosis II segregates sister chromatids into separate daughter cells*, and Figure 10.14, The Main Events in Meiosis.

9. c. Randomly removing a somatic cell from a human would give you a diploid cell with 46 chromosomes. It would also likely be in interphase, as that is the longest part of the cell cycle. For more information, see Section 10.3, *The karyotype describes all the chromosomes in a nucleus.*

10. a. Both mitosis and meiosis have similar division events in plants and animals. The main difference is cytokinesis, which must be different because of the presence of the plant cell wall. Refer to Figure 10.11, Cytokinesis in an Animal Cell; Figure 10.10, Stages of Mitosis and Cytokinesis; and Figure 10.12, Cell Division in Plants, for a diagram of this process. For more information, see Section 10.4, *The cytoplasm is divided during cytokinesis.*

11. e. Cells are diploid throughout all of mitosis and throughout most of the first division of meiosis. For more information, see Figure 10.10, Stages of Mitosis and Cytokinesis, and Figure 10.14, The Main Events in Meiosis.

12. b. The diploid number for humans is 46, so this cell must be haploid and have already undergone meiosis I. Given that its chromosomes are unduplicated, it must have completed anaphase of meiosis II when the duplicate chromatids are separated. That leaves telophase II as the most likely stage. For more information, see Section 10.5, *Meiosis II segregates sister chromatids into separate daughter cells*, and Figure 10.14, The Main Events in Meiosis.

13. e. All of the processes listed contribute to the genetic variability that exists within a sexually reproducing species such as humans. For more information, see Section 10.5, *Meiosis and fertilization contribute to genetic variation in a population, Crossing-over shuffles alleles,* and *The independent assortment of homologous pairs generates diverse gametes.*

CHAPTER 11 | Stem Cells, Cancer, and Human Health

GETTING STARTED

Below are a few questions to consider before reading Chapter 11. These questions will help guide your exploration and assist you in identifying some of the key concepts presented in this chapter.

1. How many deaths are caused by cancer each year in the United States?

2. What is the main difference between embryonic and adult stem cells? What types of cells can each stem cell differentiate into, given the appropriate conditions?

3. How do cancerous cells and tumors begin to form? How do positive and negative growth regulators work together to ensure that cells do not continue to grow in an unregulated fashion?

4. What are mutations, and what is their role in the development of cancer?

5. How is the delicate balance between proto-oncogenes and tumor suppressor genes important for cells to avoid the development of cancer?

6. What is the greatest challenge in battling cancer with treatments such as radiation therapy and chemotherapy?

7. What is the function of the enzyme telomerase? Which type of cells produce telomerase?

8. HeLa cells are considered immortal. What does "immortal" mean in relation to a cell? How can these cells be used in cancer research?

A GUIDE TO THE READING

The following concepts typically give students the most difficulty when exploring the content in Chapter 11 for the first time. For each concept, one or more references have been identified that may help you gain a better understanding of these potentially problematic areas.

Proto-Oncogenes and Tumor Suppressors

As we saw in the chapter, genes responsible for causing cancer are referred to as "oncogenes." Oncogenes typically are genes that have a normal cellular function but have mutated or become overactivated. When this occurs, the altered gene then serves to stimulate cell division, resulting in unregulated cell growth and the development of cancer. Because these genes can exist in two states, they are often called "proto-oncogenes" (normal) or "oncogenes" (altered, cancer-causing). Luckily, the cell does not rely on the health of proto-oncogenes to avert the development of cancer. Typically, there are several safeguards that help control the cell division process. These safeguards, called "tumor suppressors," are genes whose function it is to halt cell (and tumor) growth. Therefore, it is the balance between the action of proto-oncogenes and tumor suppressors that controls whether a cell continues to divide. When this system fails, the result is tumor formation.

For more information on this concept, be sure to focus on

- Section 11.2, *Gene mutations are the root cause of all cancers*
- Figure 11.10, Growth Factors Can Be Positive or Negative Regulators of the Cell Cycle

Cancer as a Multistep Process

The development of tumors can result when the cell division regulatory machinery becomes defective. As we saw in Chapter 11, this regulatory machinery involves a complex series of safeguards that help protect the cell. For a cell to become cancerous, several of these safeguard checkpoints must be overcome. A particularly good example described in the chapter is that of colon cancer. For a malignant tumor to develop in the colon, a cell in the wall of the colon must first lose the function of two or more tumor suppressor genes, have a mutation in a proto-oncogene that results in the gene becoming an oncogene, and completely lose the *p53* tumor suppressor gene. If all these events occur in the proper sequence, a small benign polyp may ultimately develop into a malignant tumor. The key to understanding this process is realizing that if any of these conditions are not met, cancer will not develop.

For more information on this concept, be sure to focus on

- Section 11.2, *Gene mutations are the root cause of all cancers*
- Section 11.2, *Cancer develops as multiple mutations accumulate in a single cell*
- Figure 11.12, Development of Colon Cancer Is a Multi-step Process

TYING IT ALL TOGETHER

Several concepts presented in this chapter build on those presented in previous chapters and may also be revisited and discussed in greater detail in subsequent chapters, including

Viruses

- Chapter 2—Section 2.4, Viruses: Nonliving Infectious Agents

Cell Communication and Signaling

- Chapter 7—Section 7.6, Cell Signaling

Human Genetics

- Chapter 13—Section 13.1, The Role of Chromosomes in Inheritance

Mutation

- Chapter 14—Section 14.4, Repairing Replication Errors and Damaged DNA
- Chapter 15—Section 15.6, The Effect of Mutations on Protein Synthesis

Control of Gene Expression

- Chapter 14—Section 14.8, How Cells Control Gene Expression

PRACTICE QUESTIONS

Factual Knowledge

1. Which of the following is *not* a difference between benign and malignant tumors?
   a. A benign tumor is a solid cell mass formed by the proliferation of cells, whereas a malignant tumor is caused by viruses.
   b. Malignant tumors may invade the bloodstream and other tissues of the body.
   c. Benign tumors can be removed surgically with relative ease.
   d. Benign tumors are often precursors to malignant tumors.
   e. All of the above are differences between benign and malignant tumors.

2. Stem cells can differentiate into different cells depending on which of the following?
   a. gene expression of parent stem cell
   b. physical environment
   c. chemical environment
   d. external stimuli
   e. all of the above

3. Which of the following statements regarding mutations is incorrect?
   a. A mutation is a change in the DNA sequence of a gene.
   b. Mutations can result in a reduction in a gene's function.
   c. Mutations can result in an increase in a gene's function.
   d. Mutations can result in the increased production of a particular protein.
   e. all of the above

4. Which of the following is *not* a typical step in cancer progression?
   a. cell proliferation
   b. metastasis
   c. tissue invasion
   d. telomerase inhibition
   e. loss of anchorage dependence

5. Most human proto-oncogenes are
   a. very uncommon and almost never mutate into oncogenes.

b. common but unlikely ever to be changed into oncogenes.

c. common and changed into oncogenes mainly by viral infections.

d. common and changed into oncogenes mainly by environmental factors.

e. always changed into the *Src* gene by viral infection.

6. One reason proto-oncogenes are especially susceptible to becoming oncogenes is that they code for the production of _____ that are part of the normal signal cascades involved in the control of cell division.
   a. polyps
   b. Rous sarcomas
   c. positive growth regulators
   d. phospholipids
   e. all of the above

7. An example of a positive growth regulator is _____, which will _____ the cell cycle, whereas a negative growth regulator such as _____ will _____ the cell cycle.
   a. EGF, inhibit, TGF-B, promote
   b. EGF, promote, TGF-B, inhibit
   c. TGF-B, inhibit, EGF, promote
   d. TGF-B, promote, EGF, inhibit
   e. EGF, inhibit, TGF-B, inhibit

8. Unlike oncogenes, both copies of a tumor suppressor gene must be mutated for a cell to lose control of its negative growth regulation. (True or False)

9. If a cell has complete control over both negative and positive regulation of its division, which of the following will most likely occur?
   a. The cell will divide rapidly but not become a tumor.
   b. The cell will divide to become a benign tumor.
   c. The cell will divide to become a malignant tumor.
   d. The cell will divide at its normal rate.
   e. No signal cascade molecules will be produced in the cell.

10. Nearly half of all cancers can be traced to purely inherited genetic defects. (True or False)

11. What factors can increase human susceptibility to cancer?
    a. exposure to hormones
    b. carcinogens
    c. obesity
    d. unprotected sex
    e. all of the above

12. In which of the following human organs do 100,000 or more new cases of cancer develop annually in the United States?
    a. breast
    b. lung
    c. prostate
    d. colon
    e. all of the above

13. Which of the following statements regarding cancer risk is true?
    a. Identical twins have a higher risk of getting cancer than nonidentical twins.
    b. Inherited genetic defects pose the most significant risk for developing cancer.
    c. Environmental factors pose the greatest cancer risk.
    d. Viruses are the only known risk for developing cancer.
    e. none of the above

14. Smoking has been linked to which of the following cancers?
    a. lung
    b. bladder
    c. kidney
    d. stomach
    e. all of the above

15. Match each term with the best description.
    ___ benign
    ___ malignant
    ___ *p53*
    ___ proto-oncogene
    ___ tumor suppressor
    a. an example of a critical tumor suppressor gene
    b. tumor that spreads to other parts of the body
    c. *Rb* protein, for example
    d. relatively harmless tumor
    e. predecessor of oncogene

Conceptual Understanding

1. Cigarette smoking has been shown to cause cancer. Which of the following statements regarding the link between smoking and cancer is *false*?
   a. Cigarette smoke contains carcinogens such as polycyclic aromatic hydrocarbons (PAHs).
   b. PAHs form physical complexes with DNA called "adducts."
   c. The *p53* gene contains several sites for adduct formation.
   d. The formation of adducts in the *p53* gene is restricted to lung cells in smokers.
   e. none of the above

2. Refer to Figure 11.4 in your textbook. The blastocyst is essentially a ball of cells that forms approximately one week from fertilization. The blastocyst has had the opportunity to form the three layers of an embryo, the endoderm, mesoderm, and ectoderm. (True or False)

3. What do *p53* and *Rb* genes have in common?
   a. They are both oncogenes.
   b. They are both proto-oncogenes.
   c. They are both tumor suppressor genes.
   d. They both promote excessive cell division.
   e. Neither is involved in regulating signal cascades.

4. For a cell to undergo the process of normal cell division,
   a. proto-oncogenes must be activated and tumor suppressors must be inactivated.
   b. proto-oncogenes must be inactivated and tumor suppressors must be activated.
   c. oncogenes must be activated and proto-oncogenes must be inactivated.
   d. oncogenes must be activated and tumor suppressors must be inactivated.
   e. none of the above

5. Stem cells are found in both embryos and adults but have the opportunity to differentiate differently. Why do cancer patients going through chemotherapy and radiation have side effects such as hair loss and extreme fatigue?
   a. Stem cells are not given the opportunity to grow and develop during these treatments.
   b. The body is using all of the supplied energy to help develop more cells because treatments kill *all* developing or dividing cells.
   c. Chemotherapy interferes with the cell division process.
   d. Depending on the cancer, radiation and chemotherapy are not always localized and thus may affect all cells in the body.
   e. All of the above.

6. Why is it more common for a person beyond middle age to have an increased risk for cancer?
   a. Older cells have fewer stem cells to help produce newer cells.
   b. Proto-oncogenes and tumor suppressor genes will have a longer time to accumulate mutations.
   c. Genes can no longer repair themselves and will fall to mutations.
   d. Hereditary mutations will occur after this age to cause cancer.
   e. Oncogenes and tumor suppressor genes reduce the number of mutations.

7. Referring to Figure 11.12, Development of Colon Cancer is a Multistep Process, determine the latest step in which a cancer treatment would offer a person a good chance to survive.
   a. step 1
   b. step 2
   c. step 3
   d. step 4
   e. Colon cancer is always fatal.

8. The vast majority of human cancer cases
   a. are colon cancer.
   b. occur in children.
   c. involve women.
   d. are caused by inherited genetic conditions.
   e. are caused in part or completely by environmental factors.

9. It usually takes only one or two mutations of proto-oncogenes to trigger the development of most cancers in humans. (True or False)

10. Which of the following represent current medical approaches to treating patients with cancer?
    a. using genetically engineered viruses that target and destroy cancer cells
    b. using drugs to disable proteins that allow cancer cells to divide indefinitely
    c. using drugs to block the growth of blood vessels in cancerous tissues
    d. surgical removal of tumor masses
    e. all of the above

11. Which of the following statements regarding growth factors is *false*?
    a. They are a type of positive growth regulator.
    b. They activate signal cascades that typically promote cell division.
    c. They typically block the activity of tumor suppressors.
    d. They can work to activate protein kinases.
    e. none of the above

12. In colon cancer, the complete loss of one tumor suppressor, combined with the mutation of a proto-oncogene, can lead to the formation of
    a. benign polyps.
    b. middle-stage polyps.
    c. late-stage polyps.
    d. a malignant tumor.
    e. none of the above

13. Both colon cancer and smoking-induced lung cancer
    a. are inherited genetic defects.
    b. are caused by an overactive proto-oncogene.
    c. involve inactivation of *p53*.

d. are caused by rare viral infections.

e. none of the above

RELATED ACTIVITIES

- Search the Internet for information on the relation of diet to colon cancer. What kinds of food are thought to increase the occurrence of colon cancer, and which to reduce it? What is known about the way diet affects the positive and negative growth regulation factors that control cell division? Compose a one-page essay summarizing the information you find.

- Search the Internet for additional information about the human cancers listed in Table 11.2 of your textbook. Discuss any environmental factors that are thought to be associated with the development of these cancers, and how these factors specifically cause the mutation of proto-oncogenes into oncogenes. Compose a one-page essay summarizing the results of your search.

- Using the chapter's Biology Matters box, "Avoiding Cancer by Avoiding Chemical Carcinogens," list at least five lifestyle changes you and your family can make to help avoid cancer. Use your new science knowledge to explain these changes to your family, and make a game out of the changes. Make a checklist for each member of your family and then see who makes the best and greatest number of changes to help avoid cancer.

ANSWERS AND EXPLANATIONS

Factual Knowledge

1. a. Recall that all tumors are essentially cell masses formed by the unchecked proliferation of cells. The type of tumor (malignant or benign) depends on the ability of the cells to migrate away from the primary tumor mass and invade other tissues in the body, as occurs with malignant tumors. For more information, see Section 11.2, Cancer Cells: Good Cells Gone Bad, particularly *Cancer develops when cells lose normal restraints on division and migration.*

2. e. All of the above. External stimuli are physical and chemical environments, and the gene expression of the parent stem cell can have an effect on the type of cell the stem cell will differentiate into. For more information, see Section 11.1, *Stem cells are a source of new cells.*

3. e. Mutations (changes in the DNA sequence of a gene) can result in an increase or decrease in the activity or production of an affected protein product. For more information, see Section 11.2, *Gene mutations are the root cause of all cancers.*

4. d. Telomerase inhibition is being researched to help stop cancer. Telomerase is produced by cancer cells to make malignant cells essentially immortal by being able to constantly proliferate. For more information, see Table 11.3 and Section 11.2, *Cancer develops when cells lose normal restraints on division and migration.*

5. d. Although some human cancers may be triggered by viruses, most are the result of numerous proto-oncogenes being mutated into oncogenes by environmental factors such as chemical pollutants. For more information, see Section 11.2, *Gene mutations are the root cause of all cancers.*

6. c. Recall that proto-oncogenes are "normal" genes that typically are involved in the positive regulation of the cell cycle. For more information, see Section 11.2, *Gene mutations are the root cause of all cancers.*

7. b. EGF is a positive growth regulator that promotes the cell cycle and TGF-B (beta) is a negative growth regulator that inhibits the cell cycle. For more information, see Figure 11.10 and Section 11.2, *Cell division is controlled by positive and negative growth regulators.*

8. True. Because the protein product made by a tumor suppressor gene has a negative influence on cell division, both copies of such a gene must be inactive to completely eliminate the negative control on cell division. For more information, see Section 11.2, *Gene mutations are the root cause of all cancers.*

9. d. This set of circumstances provides the regulatory balance that will lead to normal cell division. For more information, see Section 11.2, *Cell division is controlled by positive and negative growth regulators.*

10. False. Only 1–5 percent of all human cancers are exclusively inherited. Rather, environmental factors play a huge role in determining cancer risk, often in combination with genetic factors. For more information, see Section 11.2, *Most human cancers are not hereditary.*

11. e. All of the above. Unprotected sex can allow contraction of the human papillomavirus (HPV). Obesity is linked with breast, prostate, and colorectal cancer. A carcinogen is any physical, chemical, or biological agent that increases the risk of cancer. Exposure to hormones through meat and dairy products or prescriptions can increase the risk of cancer. For more information, see Section 11.2, *Avoiding risk factors is the key to cancer prevention.*

12. e. For more information, see Table 11.2.

13. c. As demonstrated by the Scandinavian twin study, the most significant factor for cancer risk was the environment. For more information, see Section 11.2, *Avoiding risk factors is the key to cancer prevention.*

14. e. Smoking, recognized as the leading cause of cancer in the United States, has been linked to all of these malignancies. Other human behaviors also cause cancer. For more information, see Section 11.2, *The challenge in cancer treatment is to destroy malignant cells selectively*, and the Biology Matters box, "Avoiding Cancer by Avoiding Chemical Carcinogens."

15. d. benign
    b. malignant
    a. *p53*
    e. proto-oncogene
    c. tumor suppressor
    For more information, see Section 11.2, *Cancer develops when cells lose normal restraints on division and migration* (benign, malignant), *Gene mutations are the root cause of all cancers* (proto-oncogene, tumor suppressor), and *Cancer develops as multiple mutations accumulate in a single cell* (p53).

Conceptual Understanding

1. d. Recall that the formation of adducts in the *p53* gene because of exposure to PAHs is not restricted to lung cells. White blood cells have also shown similar susceptibility that can lead to other forms of cancer, such as leukemia. For more information, see Section 11.2, *Cancer develops as multiple mutations accumulate in a single cell*.

2. False. The blastocyst has not yet had the opportunity to differentiate, and this allows the inner cell mass to be used as embryonic stem cells to differentiate into most any other cell, given the correct conditions and differentiation signals. For more information, see Figure 11.4 and Section 11.1, *Embryonic stem cells are found only in very early stages of development*.

3. c. Both *p53* and *Rb* are tumor suppressor genes that produce proteins that play a role in the development of cancer. For more information, see Section 11.2, *Cancer develops as multiple mutations accumulate in a single cell*.

4. a. For normal cell division, the promoting effects of positive controls must be "on" while the inhibitory effects of negative controls must be "off." For more information, see Section 11.2, *Cell division is controlled by positive and negative growth regulators* (to understand *how* positive and negative growth regulators affect cell division), and Section 11.2, *Gene mutations are the root cause of all cancers* (to understand what proto-oncogenes, oncogenes, and tumor suppressor genes are in relation to being positive or negative growth regulators).

5. e. All of the above. For more information, see Section 11.2, *The challenge in cancer treatment is to destroy malignant cells selectively*.

6. b. Mutations can accumulate in cells, and if tumor suppressor genes and proto-oncogenes are older, such as in a middle-aged person, they have an increased chance of accumulating such mutations, which then can cause cancer. For more information, see Section 11.2, *Most human cancers are not hereditary*.

7. c. Technically, however, treatment could be done at any of the four steps. The *p53* gene is critical in suppressing tumors and protecting cellular processes. The best chance for survival would require some type of therapy at stage 3, 2, or 1. In stage 4, the cancerous cells are involved in a malignant tumor, and several types of therapy would be required for a slight chance of survival. For more information, see Section 11.2, *Cancer develops as multiple mutations accumulate in a single cell* and *The challenge in cancer treatment is to destroy malignant cells selectively*. See also Figure 11.12.

8. e. Environmental factors play at least some role in more than 95 percent of human cancers. For more information, see Section 11.2, *Avoiding risk factors is the key to cancer prevention*.

9. False. Usually human cancers develop as the result of a series of mutations involving several proto-oncogenes. For more information, see Section 11.2, *Gene mutations are the root cause of all cancers* (for background information on proto-oncogenes) and, more specifically, *Cancer develops as multiple mutations accumulate in a single cell*.

10. e. All of the medical strategies listed are either in use or in clinical trial for treating patients with cancer. For more information, see Section 11.2, *The challenge in cancer treatment is to destroy malignant cells selectively*.

11. c. Recall that growth factors act as positive growth regulators whose activity is counterbalanced by the activity of tumor suppressors. Therefore, they have no direct influence on the activity of tumor suppressors in the cell. For more information, see Section 11.2, *Cell division is controlled by positive and negative growth regulators*.

12. b. For more information, refer to Figure 11.12. Also see Section 11.2, *Cancer develops as multiple mutations accumulate in a single cell*.

13. c. The *p53* protein plays a role in both of these types of cancer. In colon cancer, inactivation of *p53*, in conjunction with the loss of a tumor suppressor and overactivation of a proto-oncogene, can result in the formation of carcinomas. In the case of lung cancer, repeated exposure to the carcinogens present in cigarette smoke eventually inactivates the *p53* present in lung cells, leading to the development of tumors. For more information, see Section 11.2, *Cancer develops as multiple mutations accumulate in a single cell*.

CHAPTER 12 | Patterns of Inheritance

Below are a few questions to consider before reading Chapter 12. These questions will help guide your exploration and assist you in identifying some of the key concepts presented in this chapter.

1. What was the legend of Princess Anastasia, and how did the science of genetics play a part in solving this mystery?

2. What organism did the "Father of Genetics," Gregor Mendel, work with to devise his theory of inheritance?

3. What is the ultimate source of all new genetically inherited traits?

4. Why do true-breeding purple-flowered pea plants, when crossed with true-breeding white-flowered pea plants, only produce purple-flowered offspring?

5. How did the experiments of Gregor Mendel disprove the theory of blending inheritance?

6. What is a Punnett square, and how can it be used to illustrate Mendel's law of segregation?

7. What is incomplete dominance?

8. Why might a mouse with two copies of the dominant black fur gene still have white fur?

9. What are polygenic traits, and how do they determine characteristics such as skin color in humans?

A GUIDE TO THE READING

The following concepts typically give students the most difficulty when exploring the content in Chapter 12 for the first time. For each concept, one or more references have been identified that may help you gain a better understanding of these potentially problematic areas.

Dominant vs. Recessive

The concept of gene dominance is vital to the understanding of inheritance. Recall that the phenotype of an organism refers to the physical, observable trait resulting from the individual's genotype, or genetic makeup. A perfect example of this is flower color in pea plants, as studied by Mendel. In this case, flower color is determined by two different alleles, or versions, of a single gene. The P allele contains the genetic code for producing purple flowers (the phenotype). The p allele contains the genetic code for producing white flowers. When a single copy of each allele is present in the same plant, the resulting genotype of the organism is Pp. Because the two copies of the gene are dissimilar, we refer to this organism as a "heterozygote." In heterozygous organisms, the influence of one gene may take precedence over the other in the generation of the phenotype. In this example, the heterozygous plant would produce purple flowers. Therefore, the P allele is considered to be dominant to the recessive p allele because the purple flower phenotype is displayed over the white flower phenotype. Keep in mind that the terms "dominant" and "recessive" are comparative; an allele, if considered on its own, cannot be determined to be dominant or recessive unless one is comparing it to an alternate form of the same gene.

For more information on this concept, be sure to focus on

- Section 12.1, Essential Terms in Genetics
- Section 12.2, Basic Patterns of Inheritance

Mendel's Law of Segregation

As we saw in the chapter, Mendel's law of segregation states that "the two copies of a gene are separated during meiosis and end up in different gametes." To understand this statement, be sure to review the discussion of meiosis in Chapter 10 (in particular, Section 10.5, Meiosis: Halving the Chromosome Set to Make Gametes). If you understand that it is physically impossible for more than one copy of a single gene to end up in a single sperm or egg cell (in normal circumstances), then understanding the law of segregation becomes easy. Keep in mind that it is this law of segregation that makes it possible to use the Punnett square to predict the probability of obtaining offspring with a given genotype and phenotype.

For more information on this concept, be sure to focus on

- Section 12.3, *Mendel's single-trait crosses revealed the law of segregation*
- Figure 12.7, The Punnett Square Method

Mendel's Law of Independent Assortment

As we saw in the text, the law of independent assortment states that "when gametes form, the two copies of any given gene (alleles) segregate during meiosis independently of any two alleles of other genes." This law basically states that each gene present in an organism is inherited independently of all the other genes. Mendel showed this with his experiments on pea plants while observing the inheritance of multiple traits (such as pea color and shape) simultaneously. Recall that although the pea color trait (yellow vs. green) and the pea shape trait (wrinkled vs. smooth) both obey the law of segregation as expected, these two traits behave independently of each other during the process of meiosis. Therefore, the inheritance of a particular pea color has no influence whatsoever on the inheritance of pea shape.

For more information on this concept, be sure to focus on

- Section 12.3, *Mendel's two-trait experiments led to the law of independent assortment*
- Figure 12.8, Inheritance of Two Traits over Three Generations

Incomplete Dominance and Codominance

Not all traits follow a simple dominant-recessive pattern. For these traits, an allele may not produce its maximum phenotype or may contribute to multiple phenotypes. Consider a gene with two alleles that produce distinct phenotypes. When heterozygotes display a phenotype that is intermediate between the two phenotypes displayed by homozygotes, the occurrence is referred to as "incomplete dominance." The examples used in the text include flower color in snapdragons and coat color in horses. The key to identifying incomplete dominance is the fact that heterozygous individuals display a phenotype different from either of the two homozygous forms. This is in contrast to codominance, where "the phenotype of the heterozygote is determined equally by each allele." In this case, different alleles of the same gene may produce their maximum phenotype simultaneously. The example provided in the text is that of blood types in humans, where two alleles (*A* and *B*) will be expressed regardless of the combination of alleles inherited. The key to identifying codominance is understanding that heterozygous individuals essentially display two phenotypes (one for each different allele) simultaneously. In the case of blood types, an individual who is heterozygous (*AB*) will display the type *A* phenotype and the type *B* phenotype simultaneously. Blood types are complicated slightly by the presence of a third allele (*O*), which is recessive to both the *A* and the *B* alleles.

For more information on this concept, be sure to focus on

- Section 12.1, Essential Terms in Genetics
- Section 12.4, *Many alleles display incomplete dominance*
- Figure 12.10, Incomplete Dominance in Horses
- Figure 12.11, Genetic Basis of the ABO Blood Types in Humans

Epistasis

Genes are complex entities. Rarely does a gene exert its phenotypic effect without interaction from the presence or absence of other genes. When the effect of a particular allele depends on the presence of other alleles from a separate, distinct gene, this condition is called "epistasis." The example provided in the text for epistasis is coat color in mice. Separate genes for the pigment melanin and a biochemical precursor to the pigment are inherited independently. In this case, the alleles for melanin production result in either black fur (*B*, dominant) or brown fur (*b*, recessive). Both alleles for pigment production rely on the second gene for the precursor, which has two alleles as well (*C*, dominant, pigment produced; *c*, recessive, no pigment produced). Therefore, regardless of the genotype for fur color, if the mouse is homozygous recessive for the precursor (*cc*), the mouse will be incapable of producing pigment and thus will be white. In this case, the melanin gene is dependent on the independently inherited precursor gene. When considering epistasis, it is important to remember that even though multiple genes may interact to produce a single phenotype, these genes still must obey the law of independent assortment.

For more information on this concept, be sure to focus on

- Section 12.4, *Alleles for one gene can alter the effects of another gene*
- Figure 12.14, Alleles of One Gene May Affect the Phenotype Produced by Alleles of Another Gene

TYING IT ALL TOGETHER

Several concepts presented in this chapter build on those presented in previous chapters and may also be revisited and discussed in greater detail in subsequent chapters, including

Meiosis and the Production of Gametes

- Chapter 10—Section 10.5, Meiosis: Halving the Chromosome Set to Make Gametes
- Chapter 13—Section 13.1, The Role of Chromosomes in Inheritance

Sex-Linked Inheritance

- Chapter 13—Section 13.6, Sex-Linked Inheritance of Single-Gene Mutations

Exceptions to Mendelian Genetics

- Chapter 13—Section 13.3, Genetic Linkage and Crossing-Over

Mutations

- Chapter 15—Section 15.6, The Effect of Mutations on Protein Synthesis

DNA Technology and Human Disease

- Chapter 16—Section 16.5, Human Gene Therapy

PRACTICE QUESTIONS

Factual Knowledge

1. Which of the following statements regarding genes is *false*?
   a. Genes are located on chromosomes.
   b. Genes consist of a long sequence of DNA.
   c. Genes contain information for the production of a single protein.
   d. In sexually reproducing species, each cell contains a single copy of every gene.
   e. none of the above

2. Mendel used _____ as his research organism to study patterns of genetic inheritance.
   a. garden peas
   b. snapdragons
   c. horses
   d. Siamese cats
   e. mice

3. Mendel's primary contribution to our understanding of genetic inheritance was
   a. the idea that genes are found on chromosomes.
   b. providing a mechanism that explains patterns of inheritance.
   c. describing how genes are influenced by the environment.
   d. determining that DNA contains information that codes for proteins.
   e. the discovery of alleles.

4. In Mendel's model of inheritance, what he described as units of inheritance we now refer to as
   a. genes.
   b. chromosomes.
   c. homozygotes.
   d. heterozygotes.
   e. phenotypes.

5. A diploid organism that is heterozygous for a given gene has the same number of alleles as does an organism that is homozygous for that gene. (True or False)

6. A purple-flowered pea plant has the genotype *PP*. Which of the following statements about this plant is *false*?
   a. Its phenotype is white.
   b. It has a homozygous dominant genotype.
   c. When bred to a white-flowered plant, its offspring will all be purple.
   d. The gametes produced will all have the same flower color allele.
   e. It is of the true-breeding variety.

7. An individual with the genotype *ii* would have blood type
   a. A
   b. B
   c. AB
   d. O
   e. cannot determine

8. The Punnett square method is useful as a tool for analyzing genetic crosses because alleles separate equally during gamete formation. (True or False)

9. Being able to use a Punnett square to track the pattern of inheritance in a two-characteristic cross (for example, pea seed color and seed shape) demonstrates Mendel's idea of
   a. allele segregation.
   b. blending inheritance.
   c. environmental influences on genes.
   d. independent assortment.
   e. incomplete dominance.

10. Mendel's laws of segregation and independent assortment both have their biological basis in events that take place during mitosis. (True or False)

11. The observation that individuals afflicted with albinism also always have vision problems is an example of
    a. codominance.
    b. incomplete dominance.
    c. pleiotropy.
    d. epistasis.
    e. polygenesis.

12. A chestnut-colored horse is mated with a cremello (cream-colored) horse. Over a 10-year period, all of their offspring are palominos. This pattern of inheritance is best explained by
    a. complete dominance.
    b. incomplete dominance.
    c. multiple gene effects.
    d. environmental effects on genes.
    e. small sample sizes.

13. In humans, the genetic commonality of height and skin tone is that they are both
    a. regulated by the same pleiotropic gene.
    b. strictly environmentally induced with little or no genetic component.
    c. clear violations of Mendel's basic laws of genetic inheritance.
    d. controlled by multiple genes with a strong environmental influence.
    e. cases of genes exhibiting incomplete dominance.

14. Match each term with the best description.
    __ blending inheritance
    __ heterozygote
    __ recessive allele
    __ incomplete dominance
    __ phenotype
    a. observable characteristic of genes
    b. genotype with dissimilar alleles
    c. producing a heterozygote that is an intermediate in phenotype between the two homozogotes for a particular allele
    d. without phenotypic effect when paired with a dominant allele
    e. popular theory before Mendel's research

Conceptual Understanding

1. Which of the following statements about mutations is *false*?
    a. Mutations are the source of new alleles for genes.
    b. Organisms are able to create mutations to meet their specific needs.
    c. Mutations are random events and can happen in any cell at any time.
    d. Most mutations tend to be harmful or have no effect on organisms.
    e. Occasionally, mutations can be beneficial to organisms.

2. Two organisms that are true-breeding for a certain genetic characteristic are mated and their offspring analyzed. Which of the following statements about this situation is true?
    a. Both parents are homozygotes.
    b. The offspring are either 100 percent homozygotes or 100 percent heterozygotes.
    c. The offspring represent the F1 generation.
    d. The gametes produced by the offspring will carry only one allele for this gene.
    e. all of the above

3. A pea plant that is heterozygous for the flower color gene makes gametes. What is the probability that a specific gamete contains the recessive white allele for flower color?
    a. 0 percent
    b. 25 percent
    c. 50 percent
    d. 75 percent
    e. 100 percent

4. The only known violation of Mendel's law of independent assortment is flower color in snapdragons, which is best explained using the concept of blending inheritance. (True or False)

5. The variation in coat color for the three mice pictured in Figure 12.14 of your textbook is the result of a single-gene trait modified by environmental influences. (True or False)

6. Consider a gene with two alleles that show complete dominance. When two heterozygotes for this gene breed, they have a 25 percent chance of producing a homozygous recessive offspring. The next time these two individuals breed, what are the chances that they will once again have a homozygous recessive progeny?
    a. 0 percent
    b. 25 percent
    c. 50 percent
    d. 75 percent
    e. 100 percent

7. Assuming your blood type is AB, you can donate blood to people with type _____ blood and can receive blood from people with type _____.
    a. O; AB
    b. A; B

c. B; A

d. AB; O

e. O; O

8. Consider Figure 12.8. The 9:3:3:1 ratio of phenotypes that occurs in the F2 generation of this cross can be explained using

    a. blending inheritance.

    b. Mendel's law of segregation.

    c. Mendel's law of independent assortment.

    d. both of Mendel's laws of genetic inheritance.

    e. incomplete dominance.

9. Genetically identical plant clones can exhibit dramatic phenotypic variation depending on the environmental conditions under which they are grown. (True or False)

10. Theoretically, a Siamese cat raised from birth in a very cold environment should have almost no black pigment in its fur. (True or False)

11. Recall the genetic determination of coat color in mice, as we saw in your textbook. A cross between a male with the genotype *BBcc* and a female with the genotype *bbcc* will produce offspring with which coat color(s)?

    a. black only

    b. brown only

    c. white only

    d. black and brown

    e. black, brown, and white

12. In a particular plant, two genes control leaf shape and color. Round leaves (*R*) are dominant to jagged leaves (*r*). Yellow fruits (*Y*) are dominant to white fruits (*y*). A true-breeding round-leaved, yellow-fruited plant is mated with a jagged-leaved, white-fruited plant. What are the genotypes of the plants involved in this cross?

    a. *RRYY · RRYY*

    b. *RRYY · rryy*

    c. *RrYy · RrYy*

    d. *RrYy · rryy*

    e. cannot determine from the information given

13. Using the information given in question 12, what is (or are) the phenotype or phenotypes of the offspring produced by this cross?

    a. round leaves, yellow fruit

    b. round leaves, white fruit

    c. jagged leaves, yellow fruit

    d. jagged leaves, white fruit

    e. cannot determine from the information given

14. Again using the information given in question 12, what is the probability that this cross will produce any offspring with the genotype *rrYy*?

    a. 0 percent

    b. 25 percent

c. 50 percent

d. 75 percent

e. 100 percent

NOTE: For more practice with the concepts of genetic inheritance, we strongly suggest that you work through the sample genetics problems at the end of Chapter 13 in your textbook.

RELATED ACTIVITIES

- Search the Internet for three examples of the way basic mechanisms of genetic inheritance are being used to (1) produce better food, (2) breed better plants or animals for the gardening and dairy industries, and (3) fight human disease. Write a paragraph summarizing each example, and state how Mendelian principles of genetics apply.

- Many purebred dogs are particularly susceptible to genetic abnormalities. Research the topic of inbreeding in your library or on the Internet, and find an example of genetic inbreeding in dogs. Explain how this inbreeding has occurred, what the biological consequences are for the dogs themselves, and what can be done to reduce the problem.

- This chapter's Biology in the News Box, "A Family Feud over Mendel's Manuscript on the Laws of Heredity," highlights, among other things, the importance of peer review and collaboration in science. Scientists do their best work when they are part of a community where ideas can be debated and shared. Use library or Internet resources to find and learn about a key scientific discovery that was made by multiple scientists working in collaboration. (Examples include the famous partnerships of Watson and Crick or Messelson and Stahl.) Write a one-page paper on the work done by your chosen scientists, highlighting the different strengths that each of them brought to the project.

ANSWERS AND EXPLANATIONS

Factual Knowledge

1. d. Recall that in sexually reproducing species, each cell (except for gametes) contains two copies of every gene, one copy inherited from each parent. For more information, see Section 12.1, *Diploid cells have two copies of every gene.*

2. a. Mendel first used the garden pea to make his revolutionary study of inheritance patterns. For more information, see Section 12.2, *Mendel's genetic experiments began with true-breeding pea plants.*

3. b. Mendel gave us a viable mechanism for explaining inheritance. All of the other contributions were

made by scientists who did their work after Mendel. For more information, see the chapter Introduction, Humans Have Used the Principles of Inheritance.

4. a. We now know that genes (or alleles) are the units of genetic information that Mendel was describing when he imagined their segregation into gametes. For more information, see Section 12.2, *Mendel inferred that inherited traits are determined by genes,*

5. True. Because both organisms are diploid, both individuals would have the same number of alleles, just different combinations of alleles. For more information, see Section 12.1, *Genotype directs phenotype.*

6. a. According to the shorthand notation used in the textbook, *PP* is the homozygous dominant genotype that results in a purple phenotype, not white. For more information, see Section 12.1, Essential Terms in Genetics. See also Section 12.2, *Mendel's genetic experiments began with true-breeding pea plants.*

7. d. Recall that in the case of blood type in humans, the *i* allele does not produce any cell surface proteins. Therefore, for an individual that is *ii*, no cell surface proteins would be present on red blood cells and the person would have blood type O. For more information, see Section 12.4, *The alleles of some genes are codominant.* See also the chapter's Biology Matters box, "Know Your Type."

8. True. The first step in making a Punnett square is assigning the possible gametes that can be made by each parent to adjacent sides of the square, assuming that all gametes can be made with equal probability. This assumption is possible because of allele segregation during meiosis. For more information, see Section 12.3, *Mendel's single-trait crosses revealed the law of segregation.*

9. d. The law of independent assortment states that the alleles for different genes segregate (assort) independently of one another during gamete formation. This is true because those genes are likely on separate chromosomes, which do not influence each other when they separate during meiosis. For more information, see Section 12.3, *Mendel's two-trait experiments led to the law of independent assortment.*

10. False. The events responsible for the laws of segregation and independent assortment occur in meiosis, not mitosis. For more information, see Section 12.2, *Mendel inferred that inherited traits are determined by genes,*

11. c. Recall that "pleiotropy" refers to situations in which a single gene has influence over multiple, different traits. As we saw in the text, a pleiotropic gene is one that can influence two or more different traits simultaneously, such as in the example with albinism and vision problems. For more information, see Section 12.4, *A pleiotropic gene affects multiple traits.*

12. b. Because both parents are homozygotes of opposite genotypes (see Figure 12.10), their offspring must all be heterozygous for the color gene. The presence of a unique phenotype for the heterozygote suggests incomplete dominance as the mechanism of inheritance for this trait. For more information, see Section 12.4, *Many alleles display incomplete dominance.*

13. d. These traits are controlled by multiple genes. They are also strongly influenced by environmental factors, such as nutrition and sun exposure. For more information, see Section 12.4, *Most traits are determined by two or more genes.* See also Section 12.5, Complex Traits.

14. e. blending inheritance
    b. heterozygote
    d. recessive allele
    c. incomplete dominance
    a. phenotype
    For more information, see Section 12.1, Essential Terms in Genetics; Table 12.1; Section 12.2, Basic Patterns of Inheritance; and Section 12.4, *Many alleles display incomplete dominance.*

Conceptual Understanding

1. b. There is no evidence that organisms have any control over the kind of mutations that occur in their genetic material. For more information, see Section 12.1, *Gene mutations are the source of new alleles.*

2. e. Given that both parents are true-breeding, the cross must be either *AA · AA, AA · aa,* or *aa · aa.* If you work out these crosses, you will see that all of the possible answers are true for each possible situation. For more information, see Section 12.3, *Mendel's single-trait crosses revealed the law of segregation.*

3. c. Heterozygotes can make two kinds of gametes, each with equal probability. For more information, see Section 12.2, *Mendel inferred that inherited traits are determined by genes.*

4. False. Mendelian inheritance still applies in the case of flower color in snapdragons, but this phenomenon is the result of incomplete dominance. For more information, see Section 12.4, *Many alleles display incomplete dominance.*

5. False. The production of coat color in mice is an example of two genes that influence each other, with one having more prominence (epistasis). There is no environmental control for this trait. For more information, see Section 12.4, *Alleles for one gene can alter the effects of another gene.*

6. b. Each reproduction event is the result of the fusion of a single gamete from each of the two parents, and therefore they are independent events. Thus the probabilities remain the same each time. For more information, see Section 12.3, *Mendel's insights*

rested on a sound understanding of probability.

7. d. Because your blood cells would have both A and B cell surface proteins on their surface, your blood would be acceptable only to others with type AB blood. Also, people of blood type O are considered universal donors because their blood cells do not contain any cell surface proteins; and therefore you would be able to accept this blood type. For more information, see the Biology Matters box, "Know Your Type."

8. d. Because two characters are involved in this cross, both segregation and independent assortment apply. For more information, see Section 12.3, *Mendel's two-trait experiments led to the law of independent assortment.*

9. True. This is a classic example of gene and environment interactions. For more information, see Section 12.4, *The environment can alter the effects of a gene.*

10. False. Because pigment production and deposition in Siamese cats is under the control of a temperature-sensitive enzyme that works better at colder temperatures, just the opposite of this statement would be true. For more information, see Section 12.4, *The environment can alter the effects of a gene.*

11. c. Because both individuals are homozygous recessive for the *c* gene, their offspring will also be homozygous recessive for *c*. As *c* has more influence over pigmentation than does the *B* gene, their offspring will not deposit coat pigment and will therefore all be white. For more information, see Section 12.4, *Alleles for one gene can alter the effects of another gene.*

12. b. "True-breeding" in this case means homozygous for both traits, with the parents of opposite phenotypes. For more information, see Section 12.2, *Mendel's genetic experiments began with true-breeding pea plants.*

13. a. The offspring will be heterozygotes for both traits, which means that they will express both dominant phenotypes. For more information, see Section 12.2, *Mendel inferred that inherited traits are determined by genes.*

14. a. In this F_1 genotype, the leaf shape gene is homozygous recessive. This means that both parents would have had to contribute an *r* allele. Because the true-breeding, round-leaved parent does not have such an allele, this F_1 genotype is not possible through the cross described. For more information, see Section 12.2, *Mendel inferred that inherited traits are determined by genes.*

CHAPTER 13 | Chromosomes and Human Genetics

GETTING STARTED

Below are a few questions to consider before reading Chapter 13. These questions will help guide your exploration and assist you in identifying some of the key concepts presented in this chapter.

1. What is Huntington disease, and what is the significance of this disease in the field of human genetics?

2. How many different genes make up the human genome?

3. On which chromosome is the *SRY* gene located, and what role does it play in determining sex in humans?

4. How does the process of crossing-over work to generate nonparental genotypes?

5. Why are genetic disorders that are caused by recessive alleles much more common than those caused by dominant alleles?

6. What are the processes amniocentesis, chorionic villus sampling, and preimplantation genetic diagnosis used for?

7. What do the genetic disorders Down syndrome, Klinefelter syndrome, and Turner syndrome all have in common?

A GUIDE TO THE READING

The following concepts typically give students the most difficulty when exploring the content in Chapter 13 for the first time. For each concept, one or more references have been identified that may help you gain a better understanding of these potentially problematic areas.

Homologous Chromosomes

As we saw in the text, the work of Gregor Mendel and August Weismann suggested that genes were located on physical structures (later determined to be chromosomes) present in the cell. This led to the development of the "chromosome theory of inheritance." The process of inheritance involves the contribution of chromosomes from both parents. Specifically, each individual possesses one complete set of chromosomes from the father and one complete set from the mother. This results in the presence of two copies of each chromosome (and therefore two copies of each gene) in each and every cell. The pairs of chromosomes containing the same genes are referred to as "homologous chromosomes." During the process of meiosis (Chapter 10), these homologous chromosomes pair at the metaphase plate for proper independent assortment.

For more information on this concept, be sure to focus on

- Section 13.1, The Role of Chromosomes in Inheritance
- Section 13.2, Origins of Genetic Differences between Individuals

Genetic Linkage and Crossing-Over

Despite Gregor Mendel's elaborate work, scientists often observed that certain genes seemed to be inherited together, violating the law of independent assortment. This contradictory observation remained a mystery until the early 1900s, when the studies of Thomas Hunt Morgan helped solve the puzzle. Morgan, working with inherited traits in fruit flies, was able to determine that two genes physically located near one another on a chromosome were more likely to be inherited together, violating the law of independent assortment. Of course, this makes sense: we now know that chromosomes are inherited as intact units. Therefore, any genes located on the same chromosome should be inherited together in a linked fashion. However, Morgan also observed that genes that should have been linked would, on occasion, assort independently, producing unexpected phenotypes. This could be explained by the process of crossing-over, in which genetic material is exchanged between paired homologous chromosomes during meiosis. The key to understanding crossing-over is to realize that the farther apart two genes are physically located on a chromosome, the more likely it is that crossing-over will occur between these genes. Conversely, the closer two genes are located along a chromosome, the less likely it is that crossing-over will occur between them, resulting in a "tighter" genetic linkage.

For more information on this concept, be sure to focus on

- Section 13.3, Genetic Linkage and Crossing-Over
- Figure 13.6, Genes on the Same Chromosome May Not Assort Independently
- Figure 13.7, Without Crossing-Over, Genes on the Same Chromosome Would Be Completely Linked

Sex Linkage

Humans possess 23 pairs of homologous chromosomes. Twenty-two of these pairs provide both males and females with two copies of each gene. The remaining pair of chromosomes constitute the sex chromosomes (X and Y). Genes located on these chromosomes are considered to be sex-linked, because males (XY) would possess only a single copy of each of these genes. As a result, males are particularly susceptible to sex-linked recessive disorders, as the inheritance of only a single copy of the defective allele from the mother would result in the presence of the disease. This is in contrast to females, where two copies of the defective allele (one located on each X chromosome contributed by each parent) would be required. Keep in mind that 15 of the 1,200 genes found on the sex chromosomes are, in fact, shared between the X and Y chromosomes. As a result, because two copies of these 15 genes are present in both males and females, they would not be considered sex-linked.

For more information on this concept, be sure to focus on

- Section 13.6, Sex-Linked Inheritance of Single-Gene Mutations
- Figure 13.12, X-Linked Recessive Conditions Tend to Be More Common in Males than in Females

TYING IT ALL TOGETHER

Several concepts presented in this chapter build on those presented in previous chapters and are revisited and discussed in greater detail in subsequent chapters, including

Meiosis

- Chapter 10—Section 10.5, Meiosis: Halving the Chromosome Set to Make Gametes

Mendelian Genetics

- Chapter 12—Section 12.2, Basic Patterns of Inheritance
- Chapter 12—Section 12.3, Mendel's Laws of Inheritance

Noninherited Genetic Disorders

- Chapter 11—Section 11.2, *Gene mutations are the root cause of all cancers*

Sickle-Cell Anemia

- Chapter 15—Section 15.7, Putting It All Together: From Gene to Protein to Phenotype

Gene Expression

- Chapter 14—Section 14.7, Patterns of Gene Expression

DNA Technology

- Chapter 16—DNA Technology

Inheritance of Alleles and Evolution

- Chapter 18—Section 18.3, Mutation: The Source of Genetic Variation
- Chapter 18—Section 18.6, Natural Selection: The Effects of Advantageous Alleles

PRACTICE QUESTIONS

Factual Knowledge

1. Which of the following ideas or events related to the chromosomal theory of inheritance is *false*?
   a. Chromosomes were first observed in dividing cells in 1882.
   b. Mendel was the first to discover that chromosomes contain genes.
   c. Chromosomes are made of a single molecule of DNA and many proteins.
   d. Genes are located in specific places on the chromosome, called "loci."
   e. Homologous chromosomes separate into different gametes during meiosis.

2. The X chromosome in humans is
   a. an example of an autosome.

b. the only human sex chromosome.

c. present only in females.

d. always found in single copy.

e. none of the above

3. In humans, the female parent's genetic contribution determines the sex of a child. (True or False)

4. A person is genetically XX and develops as a male. How can this be explained?

a. In humans, males are XX.

b. A mistake of genetic analysis must have taken place, because XX cannot develop into a male.

c. The *SRY* gene is also present in this individual.

d. The X chromosomes have nothing to do with sex determination in humans.

e. none of the above

5. In the absence of crossing-over, Mendel's law of independent assortment does *not* apply when genes are located on the same chromosome. (True or False)

6. On chromosomes, a pair of linked genes will be found only on

a. autosomes.

b. sex chromosomes.

c. the same chromosome.

d. opposite ends of the same chromosome.

e. none of the above

7. Genes located on the same chromosome tend to be inherited as a group. Which of the following can disrupt this pattern of inheritance?

a. sex determination

b. crossing-over

c. chromosome replication

d. changes in chromosome number

e. all of the above

8. _____ is a process that contributes to genetic variation in organisms.

a. Crossing-over

b. Independent assortment

c. Gene mutation

d. Fertilization

e. all of the above

9. Refer to the human pedigree shown in Figure 13.9 of your textbook. The individual labeled "1" in generation III is

a. a healthy male.

b. a healthy female.

c. a male with brachydactyly.

d. a female with brachydactyly.

e. one of the parents.

10. Because humans have long generation times, consciously choose their mates, and produce relatively few offspring, the experimental study of genetic disorders is relatively easy. (True or False)

11. If two parents are heterozygous for an autosomal recessive disease,

a. they are both considered genetic carriers for the disease.

b. their children have no chance of inheriting the disease.

c. their children have a 50 percent chance of inheriting the disease.

d. all of their children will also be heterozygotes.

e. all of their children will have the disease.

12. In humans, X-linked genetic diseases

a. are associated with one of the sex chromosomes.

b. include hemophilia and Duchenne muscular dystrophy.

c. can be seen in both males and females.

d. tend to be expressed more in males than in females.

e. all of the above

13. Down syndrome in humans is an example of

a. normal genetic recombination.

b. a change in the overall number of chromosomes.

c. an X-linked recessive disorder.

d. a single-gene autosomal mutation.

e. a change in the structure of one of the sex chromosomes.

14. Match each term with the best description.

___ translocation

___ crossing-over

___ locus

___ pedigree

___ X-linked

a. place where a gene is found on a chromosome

b. movement of a gene to a nonhomologous chromosome

c. trait associated with one of the sex chromosomes

d. what charts genetic relationships within a family history

e. exchange of genes between homologous chromosomes

Conceptual Understanding

1. Which of the following statements about homologous chromosomes is *false*?

a. There are 22 pairs in humans, not counting the sex chromosomes.

b. They contain the same genes in the same locations.

c. They pair during meiosis.

d. They sometimes engage in crossing-over during meiosis.

e. none of the above

2. All humans are capable of producing gametes containing different sex chromosomes. (True or False)

3. Which of the following diseases is *not* influenced by a person's genetic makeup?
   a. heart disease
   b. cancer
   c. diabetes
   d. arthritis
   e. none of the above

4. Two genes are known to be on the same chromosome, yet analysis of genetic crosses involving these genes suggests that they assort independently. The most plausible explanation for this observation is that they are
   a. the genes involved in causing Huntington disease.
   b. genes that are especially prone to mutation.
   c. on opposite ends of the same chromosome.
   d. located adjacent to one another on the same chromosome.
   e. X-linked genes.

5. Crossing-over is expected to occur infrequently for genes located very close together on the same chromosome. (True or False)

6. Refer to Figure 13.6. Which result indicates that the genes for body color and wing length in fruit flies are linked?
   a. Four different phenotypes are present in the offspring.
   b. Parental genotypes are present in the offspring.
   c. Nonparental genotypes are present in the offspring.
   d. An unexpectedly low percentage of nonparental genotypes is present in the offspring.
   e. all of the above

7. Which of the following statements regarding crossing-over is *false*?
   a. Crossing-over disrupts the linkage between genes on the same chromosome.
   b. Crossing-over disrupts the linkage between genes on different chromosomes.
   c. Crossing-over produces new genetic combinations.
   d. Crossing-over produces nonparental chromosomes.
   e. none of the above

8. Consider a species that reproduces sexually and has a total of three pairs of chromosomes (two pairs of autosomes and one pair of sex chromosomes). In the absence of crossing-over, how many different offspring combinations can a mating pair produce?
   a. 4
   b. 8
   c. 9
   d. 16
   e. 64

9. Consider the pedigree for a family with brachydactyly, shown in Figure 13.9. On what basis can we conclude that the allele that causes this condition is likely dominant?
   a. No one in the entire pedigree has the condition.
   b. Most of the offspring in generation II have the condition.
   c. None of the affected individuals have unaffected parents.
   d. Only one of the males in generation III has the condition.
   e. Two-thirds of those with the condition in generation III are females.

10. Sickle-cell anemia is an inherited chronic blood disease caused by an autosomal recessive allele. Suppose a man who is homozygous recessive for the sickle-cell gene fathers a child by a woman who is a carrier for sickle-cell. What are the chances their children will exhibit the disease? (Note: A Punnett square may be useful.)
    a. 0 percent
    b. 2 percent
    c. 50 percent
    d. 75 percent
    e. 100 percent

11. Autosomal dominant diseases are exhibited by anyone who carries at least one dominant allele for that gene. How is it that dominant lethal genes, such as the one that causes Huntington disease, can persist in a population?
    a. The disease-causing allele can "hide" in the heterozygous condition.
    b. The disease develops only under the influence of other genes.
    c. The environment plays a large role in determining whether the gene is expressed.
    d. These diseases usually take effect later in life after people have had children.
    e. all of the above

12. Consider the disorder congenital generalized hypertrichosis (CGH) (Figure 13.13). What can you determine about the potential offspring of a couple in which the father has CGH and the mother does *not*?
    a. All sons will have the disorder.
    b. Half of the sons will have the disorder.
    c. All daughters will have the disorder.
    d. Half of the daughters will have the disorder.
    e. We cannot determine the outcome without knowing the genotype of the mother.

13. Cri du chat syndrome (Figure 13.15) is caused by which of the following structural changes to chromosomes?
    a. deletion

b. duplication
c. inversion
d. translocation
e. trisomy

NOTE: For more practice using the concepts of human genetics, we strongly encourage you to work through the sample genetics problems at the end of Chapter 13 in your textbook.

RELATED ACTIVITIES

- Imagine that you are in the role of having to counsel a teenager who has just learned that Huntington disease runs in her family. Describe how you would explain the biology of the disease to this person, including the genetics determining her chances of acquiring the disease and chances of passing on the disease. Would you encourage her to undergo genetic screening to find out if she has the defective gene? Why or why not?

- Using the Internet or resources from your library, research the basis of an X-linked recessive disorder in humans. Possibilities include color blindness, hemophilia, and male pattern baldness, but there are many others. Compose a one-page summary that explains how the disorder manifests itself, and provide a hypothetical pedigree to chart the inheritance of the disorder through at least three generations of a family.

- Consider the discussion on genetic testing presented in the chapter's Biology in the News box, "Stanford Students See What's in Their Genes." The article raises many interesting ethical concerns that have arisen because of the availability of genetic screening. Conduct an Internet search for a genetic disorder that runs in your family or with which you are familiar. Is there currently a genetic test for this disorder? Compose a one-page summary of what the test detects and the societal impact the availability of the test has had. In your summary, be sure to offer your own views on how this test should, or shouldn't, be used.

ANSWERS AND EXPLANATIONS

Factual Knowledge

1. b. The chromosomal theory of inheritance, including the discovery that genes are on chromosomes, was developed long after Mendel had died. For more information, see Section 13.1, The Role of Chromosomes in Inheritance.

2. e. Both X and Y are types of sex chromosomes in humans, and X is found in both males and females. Human males are XY, and females are XX. For more information, see Section 13.1, *Autosomes differ from sex chromosomes.*

3. False. The human male produces two different kinds of gametes with respect to the sex chromosomes (X or Y), whereas the female produces one (X only). It is therefore the male who determines the gender of offspring in our species. For more information, see Section 13.1, *In humans, maleness is specified by the Y chromosome.*

4. c. It is possible (although rare) for the portion of a Y chromosome that determines maleness (the *SRY* gene) to become associated with an X chromosome and thus produce male characteristics in an individual who does not have a Y chromosome. For more information, see Section 13.1, *In humans, maleness is specified by the Y chromosome.*

5. True. Linked genes (those present on the same chromosome) do not show independent assortment, unless they are located very far apart from one another on the chromosome so that crossing-over is likely to occur. For more information, see Section 13.3, *Linked genes are located on the same chromosome* and *Crossing-over reduces genetic linkage.*

6. c. Genetically linked genes are present on the same chromosome, regardless of type. For more information, see Section 13.3, *Linked genes are located on the same chromosome.*

7. b. Recall that the process of crossing-over can switch alleles between homologous chromosomes. For more information, see Section 13.3, *Crossing-over reduces genetic linkage.*

8. e. All of these processes would contribute to the generation of genetic differences between individuals. For more information, see Section 13.2, Origins of Genetic Differences between Individuals.

9. d. A dark circle indicates an affected female in this pedigree. Refer to the key that accompanies Figure 13.9. For more information, see Section 13.4, *Pedigrees are a useful way to study human genetic disorders.*

10. False. Because of these factors, as well as matters of ethical concern, it is impossible to do controlled "breeding" of humans for purposes of genetic analysis. Pedigrees are the best way to study inheritance in our species. For more information, see Section 13.4, Human Genetic Disorders.

11. a. Heterozygotes have one dominant (in this case, normal) and one recessive (in this case, disease-causing) allele. Therefore, they do not show the disease, but they can pass along the disease-causing allele to their offspring, making them genetic carriers. If two heterozygotes mate, on average one-fourth of their children will have the homozygous recessive genotype and consequently be affected by the disease. For more information, see Section 13.5, *Autosomal recessive genetic disorders are common.*

12. e. Because X chromosomes are present in males and females, both can show X-linked diseases. Most X-linked genetic diseases are recessive in nature. Males express them more frequently, however, because they need only one copy of the defective gene, instead of two for females. For more information, see Section 13.6, Sex-Linked Inheritance of Single-Gene Mutations.

13. b. Down syndrome results when an individual receives three copies of chromosome 21 as the result of an error during meiosis. For more information, see Section 13.7, *Changes in chromosome number are often fatal.*

14. b. translocation
    e. crossing-over
    a. locus
    d. pedigree
    c. X-linked
    For more information, see Section 13.1, *Genes are located on chromosomes*; Section 13.2, Origins of Genetic Differences between Individuals; Section 13.4, *Pedigrees are a useful way to study human genetic disorders*; Section 13.6, Sex-Linked Inheritance of Single-Gene Mutations; and Section 13.7, *The structure of chromosomes can change in several ways.*

Conceptual Understanding

1. e. All of the statements are true regarding homologous chromosomes. For more information, see Section 13.1, *Genes are located on chromosomes*, and Section 13.2, Origins of Genetic Differences between Individuals.

2. False. Human males are XY with respect to the sex chromosomes, whereas females are XX. For more information, see Section 13.1, *Autosomes differ from sex chromosomes.*

3. e. Many diseases, including all of those listed here, are caused by multiple gene interactions with various environmental factors. Therefore, to some degree, all of the diseases listed are influenced by a person's genetic makeup. For more information, see the chapter's Biology Matters box, "Most Chronic Diseases Are Complex Traits."

4. c. Even though genes located on the same chromosome are technically linked, genes present at opposite ends of the same chromosome would be so far apart that they have a high probability of crossing-over and therefore behaving as if on separate chromosomes. For more information, see Section 13.3, *Crossing-over reduces genetic linkage.*

5. True. The closer together two genes are, the less likely they are to be involved in a crossing-over event. For more information, see Section 13.3, *Crossing-over reduces genetic linkage.*

6. d. If these genes were not linked, we would expect a much higher percentage of nonparental genotypes than those observed. For more information, see Section 13.3, *Linked genes are located on the same chromosome.*

7. b. Recall that crossing-over occurs only between homologous chromosomes and that linked genes are found only on the same physical chromosome. Therefore, these processes do not affect genes found on different chromosomes. For more information, see Section 13.2, Origins of Genetic Differences between Individuals.

8. e. Each parent would be able to produce gametes with $2^3 = 8$ different combinations of chromosomes. In a mating event, there would therefore be $8 \times 8 = 64$ possible combinations of offspring. For more information, see Section 13.2, Origins of Genetic Differences between Individuals.

9. c. If the allele were recessive, it would be masked in carriers, individuals who are unaffected but pass the condition to their offspring. A dominant allele cannot be masked in this way, so affected individuals will always have at least one parent who is affected. For more information, see Section 13.5, *Serious dominant genetic disorders are less common.*

10. c. The cross is $Ss \times ss$, which yields the same two genotypes in the offspring. Half will be carriers; the other half will have sickle-cell disease. For more information, see Section 13.5, *Autosomal recessive genetic disorders are common.*

11. d. Late onset of the disease, after breeding has occurred, explains why people pass along this otherwise lethal mutation. For more information, see Section 13.5, *Serious dominant genetic disorders are less common.*

12. c. CGH is an X-linked, dominant trait. Because the father has CGH, this means he must have the affected X chromosome. Because all of his daughters will receive his X chromosome, they will be affected, whereas none of his sons will receive his X chromosome and therefore will be unaffected. For more information, see Section 13.6, Sex-Linked Inheritance of Single-Gene Mutations.

13. a. Cri du chat results from the loss of a critical portion of chromosome number 5. This type of event is referred to as a "deletion" (see Figure 13.14). For more information, see Section 13.7, *The structure of chromosomes can change in several ways.*

CHAPTER 14 | DNA and Genes

GETTING STARTED

Below are a few questions to consider before reading Chapter 14. These questions will help guide your exploration and assist you in identifying some of the key concepts presented in this chapter.

1. Is it possible that the legendary single-eyed Greek adversary Cyclops actually existed?

2. In the early 1900s, what molecule did most geneticists believe genes were composed of?

3. How did the work of Frederick Griffith contribute to our understanding of the molecular basis of genetic material?

4. In terms of genetics, what is transformation?

5. Who was Rosalind Franklin, and what was her contribution to the study of DNA?

6. In terms of the DNA molecule, what is a complementary strand?

7. How is the process of DNA replication similar to word processing on a computer?

8. How many rads of radiation energy are required to mortally injure a human?

9. Why must people with xeroderma pigmentosum (XP) stay out of the sun?

10. What are transposons? Where did they come from, and how much of the human genome is made up of transposons?

11. If the total amount of DNA present in your body were to be stretched end to end, how long would these molecules be, and how does this compare to the distance between the sun and Earth?

12. How many different types of cells are present in the human body? How many of these types produce the protein hemoglobin?

13. In the bacterium *E. coli*, how is the production of the amino acid tryptophan regulated?

A GUIDE TO THE READING

The following concepts typically give students the most difficulty when exploring the content in Chapter 14 for the first time. For each concept, one or more references have been identified that may help you gain a better understanding of these potentially problematic areas.

DNA Replication

The key to deciphering how DNA replication occurs came with the discovery of the three-dimensional structure of the DNA molecule. Recall that the DNA molecule consists of a pair of DNA strands, each strand complementary to the other. The two strands are held together through hydrogen bonding between the complementary bases. These hydrogen bonds form a base pair. For replication to proceed, the two strands must be separated, breaking the hydrogen bonds in the process. Once separate, each of the two strands then contains the required information to serve as a template for the construction of a new complementary DNA strand.

For more information on this concept, be sure to focus on

• Section 14.2, *DNA is built from two helically wound polynucleotides*

DNA Repair

DNA repair is vital to the survival of all organisms. The need for an efficient DNA repair mechanism is due to the many varied ways in which DNA may become damaged. As we saw in the chapter, this can occur when simple mistakes are made during the process of DNA replication or as the result of physical damage caused by radiation or chemicals. In all cases, whenever damage occurs, the cell must recognize the damage, act to remove the damaged segment of DNA, and work to replace it with the correct information. This involves the action of several specialized enzymes, all working together to maintain the accuracy of the genetic information.

For more information on this concept, be sure to focus on

Transposons

The human genome consists of approximately 3.3 billion base pairs containing roughly 25,000 genes. It is estimated that only 1.5 percent of the genome contains information encoding a protein. In addition to noncoding and spacer DNA sequences, a typical eukaryotic genome also contains a large number of transposons. Transposons are sequences of DNA that are capable of changing their position within the genome. As such, they are often referred to as "jumping genes." As we have seen in the text, it is estimated that upwards of 36 percent of the human genome is made up of transposons. Scientists believe that transposons likely originated from a viral source. How do these jumping genes change position within the genome? In many cases, the gene sequence present in the transposon codes for an enzyme that facilitates the removal of the transposon from its original location and subsequent insertion in its new location. The specialized enzymes used in the repair of DNA are capable of performing similar removal and insertion functions.

For more information on this concept, be sure to focus on

Differential Gene Expression

Humans have 220 different types of cells. Each of these cells, regardless of type, contains the exact same set of genes. So how does the body develop more than 200 different cell types even though the DNA present within each cell type is exactly the same? This is accomplished through the selective expression of a subset of genes required for each particular cell type. For example, skin and hair cells produce a strong, fibrous protein called "keratin" that allows these cells to perform their protective function. As mentioned in the chapter, red blood cells in the body are the exclusive producers of the oxygen-transporting protein hemoglobin. Examples such as these illustrate how cells become specialized depending on the subset of genes they express. In addition to this differential gene expression, a subset of genes called "housekeeping genes" is expressed in almost all cells in the body. These genes code for universally required products such as different types of RNA.

For more information on this concept, be sure to focus on

Controlling Gene Expression

As we have seen in the chapter, the cell uses several distinct methods for controlling the expression of genes. The primary mechanism utilized by most cells, however, is control over the process of transcription. Regulatory DNA sequences may be used by the cell to either promote or inhibit the transcription of particular genes. Inhibition of the expression of a gene sequence can be accomplished through the binding of a repressor protein—such as that used by the tryptophan operator discussed in the chapter—to the regulatory sequence. The presence of the repressor protein can then block the binding of RNA polymerase and subsequently inhibit the process of transcription. Conversely, promoter proteins that promote the attachment of RNA polymerase may bind to regulatory sequences. Increased binding of RNA polymerase results in the enhanced transcription of the genes. To add more complexity to the process of gene regulation, it is important to understand that the activity of these promoter or repressor proteins may be turned on or off depending on cellular or environmental conditions. Therefore, the regulation of gene expression can be thought of as an elegantly complex cascade of events that work together to make the production of proteins a highly regulated and efficient process.

For more information on this concept, be sure to focus on

- Figure 14.13, Repressor Proteins Turn Genes Off
- Figure 14.14, Steps at Which Gene Expression Can Be Regulated in Eukaryotes

TYING IT ALL TOGETHER

Several concepts presented in this chapter build on those presented in previous chapters and may also be revisited and discussed in greater detail in subsequent chapters, including

Nucleotides

- Chapter 5—Section 5.10, Nucleotides and Nucleic Acids

Cancer

- Chapter 11—Section 11.2, *Gene mutations are the root cause of all cancers*

Chromosomes

- Chapter 13—Section 13.1, The Role of Chromosomes in Inheritance

Inherited Genetic Disorders

- Chapter 13—Section 13.4, Human Genetic Disorders

The Genetic Code

- Chapter 15—Section 15.2, How Genes Guide the Manufacture of Proteins
- Chapter 15—Section 15.4, The Genetic Code

PRACTICE QUESTIONS

Factual Knowledge

1. During the first half of the twentieth century, as geneticists began the search for what genes are made of, they knew that the hereditary material
   a. was contained in cells and was passed from one generation to the next.
   b. must contain all of the information necessary for life.
   c. had to be composed of a material that could be accurately replicated.
   d. needed to be sufficiently variable to accommodate all life-forms.
   e. all of the above

2. From the very start, DNA was thought to be a good candidate for the hereditary material because geneticists believed it was a complex molecule. (True or False)

3. The chemical bonds that hold the two complementary strands of the DNA double-helix close to one another are covalent bonds. (True or False)

4. Which of the following parts of a DNA molecule are held together by hydrogen bonds?
   a. the carbons within the sugar-phosphate group
   b. the carbons within the nitrogen-containing bases
   c. nucleotide bases on opposite strands of the double helix
   d. successive nucleotides within a single strand of the double helix
   e. all of the above

5. Consider the three-dimensional structure of DNA. Which of the following has a structure analogous to a molecule of DNA?
   a. railroad tracks
   b. painted lines on a highway
   c. spiral staircase
   d. concrete sidewalk
   e. curly drinking straw

6. Even though you and your brothers or sisters are all humans, odds are that each of you has a somewhat different genetic makeup. The molecular basis for this inherited genetic variation
   a. is related to the difference in the DNA nucleotide sequence of your genes.
   b. depends on whether protein or DNA is used as the hereditary material.
   c. depends on whether adenine pairs with thymine or guanine.
   d. is the result of errors in the DNA replication process.
   e. has nothing to do with DNA.

7. When a molecule of DNA replicates without error, each of the resulting molecules contains
   a. the same amount of A as T.
   b. the same amount of G as C.
   c. one new strand and one old strand.
   d. the same sequence of bases as did the original molecule.
   e. all of the above

8. Which of the following statements about the process of DNA replication is *false*?
   a. Many different enzymes are needed for the process to function properly.
   b. Mistakes can be corrected at multiple steps in the process.

c. Uncorrected mistakes introduce mutations into the DNA base sequence.

d. Mistakes in the copying process are very common occurrences.

e. Mutations represent a source of new alleles.

9. The combined length of all of the DNA present in a single cell of your body is approximately
   a. 2 meters.
   b. the length of a football field.
   c. 10 kilometers.
   d. 100 kilometers.
   e. over 100 times the distance from Earth to the sun.

10. Which of the following would you expect to find in a sample of DNA from one of your liver cells?
    a. transposons
    b. spacer DNA
    c. DNA that codes for proteins
    d. DNA that codes for RNA
    e. all of the above

11. In eukaryotic cells, histones
    a. are the equivalent of "jumping genes."
    b. are sequences of DNA that separate genes.
    c. serve as structures around which DNA is spooled for packaging.
    d. are sequences of DNA that are removed following transcription.
    e. regulate gene expression.

12. In bacteria, genes are typically turned on and off in direct response to short-term changes caused by environmental conditions. (True or False)

13. Which of the following represents a type of house-keeping gene that would likely be expressed in a human pancreas cell?
    a. ribosomal RNA (rRNA) gene
    b. hemoglobin gene
    c. insulin gene
    d. crystallin protein gene
    e. transposons

14. The most common way gene expression is regulated in both prokaryotes and eukaryotes is through the
    a. control of messenger RNA (mRNA) translation.
    b. breakdown of proteins formed by translation.
    c. prevention of DNA uncoiling prior to transcription.
    d. control of gene transcription.
    e. none of the above

15. Match each term with the best description.
    ___ DNA bases
    ___ DNA replication
    ___ mutation
    ___ operator
    ___ genome

a. specific change in DNA base sequence
b. adenine, thymine, cytosine, guanine
c. controls transcription of a gene or group of genes
d. mechanism for copying genes
e. all the DNA in an organism, including its genes

Conceptual Understanding

1. Assume a certain molecule of DNA is composed of exactly 22 percent adenine. How much cytosine would you expect to find in this molecule?
   a. 22 percent
   b. 44 percent
   c. 56 percent
   d. 28 percent
   e. 24 percent

2. A short sequence of bases on one strand of DNA is AGTCTACCGATAGT. If this sequence serves as a template for the formation of a new strand of DNA, what will be the corresponding base sequence in the new strand?
   a. AGTCTACCGATAGT
   b. TCAGATGGCTATCA
   c. TGATAGCCATCTGA
   d. GACATCGATTCGAT
   e. THISISNOTCORRECT

3. A certain enzyme is known to be responsible for separating the two strands of a DNA molecule to facilitate the process of DNA replication. This enzyme most likely
   a. breaks covalent bonds within nucleotides.
   b. causes the DNA to twist into a helix.
   c. forms covalent bonds between nucleotides.
   d. adds a sugar-phosphate group to nucleotides.
   e. breaks hydrogen bonds between bases.

4. Assume that you chemically label both strands within a molecule of DNA. You then allow this DNA to replicate using unlabeled nucleotides. Which of the following statements about the two resulting DNA molecules would be *false*?
   a. Both molecules would contain the chemical label.
   b. One molecule would contain the chemical label, the other would not.
   c. One strand within each molecule would contain the chemical label.
   d. Assuming no replication errors, both molecules would be genetically identical.
   e. none of the above

5. An uncorrected mismatch error during human DNA replication can result in
   a. the potential for a disease or genetic disorder.
   b. the combination of two or more alleles.

c. a disruption in the double-helix structure of the DNA molecule.
d. the introduction of many new mismatch errors.
e. all of the above

6. Before damaged DNA can be repaired, which of the following must occur?
a. The damaged area must be recognized.
b. The damaged area must be removed.
c. The damaged area must be replaced.
d. A collection of various repair enzymes must be present.
e. all of the above

7. The genetic disorder XP is a tragic example of the failure of DNA repair mechanisms. Specifically, XP patients possess a _____ allele that produces a nonfunctional protein whose job it would normally be to repair DNA damage caused by _____.
a. dominant; ultraviolet light
b. dominant; arsenic
c. recessive; ultraviolet light
d. recessive; polycyclic aromatic hydrocarbons (PAHs)
e. dominant; cancer

8. Assume a cell present in your body has performed a mismatch error during the process of DNA replication. This error has resulted in the generation of a new allele for one of your genes. For this new allele to be inherited by your offspring, which of the following conditions must be true?
a. The error must be repaired.
b. The error must be caused by a chemical mutagen.
c. The error must occur in the germ line cells responsible for the production of gametes.
d. The new allele produced must be dominant.
e. The new allele produced must be recessive.

9. Refer to Figure 14.10 in your textbook. The structure labeled "metaphase chromosome" at the top of the diagram would not be able to undergo successful gene transcription because
a. there is no operator present.
b. the structure lacks regulatory genes.
c. the DNA is packed too tightly.
d. there are no housekeeping genes in this structure.
e. an active repressor is present to prevent transcription.

10. A sample of DNA from an unknown organism is analyzed and found to contain histone proteins, more than 2 billion base pairs, and large segments of noncoding DNA. From this information, one can conclude that the organism is
a. a bacterium.
b. an animal.

c. a plant.
d. some kind of eukaryotic organism.
e. none of the above

11. Which of the following would *not* be considered a mechanism used by eukaryotes to regulate gene expression?
a. tight packaging of DNA
b. regulation of transcription
c. regulation of the breakdown of mRNA molecules
d. regulation of the life span of a protein product
e. binding of tryptophan to a repressor protein

12. In which of the following cells from a human would you expect to find the genes that code for both insulin and rRNA expressed at the same time?
a. red blood cell
b. eye lens cell
c. pancreatic cell
d. white blood cell
e. all of the above

13. An *E. coli* bacterium is lacking in the amino acid tryptophan. Which of the following would be true about this bacterium?
a. The tryptophan-repressor protein complex is activated.
b. The tryptophan-repressor protein complex can bind to the operator.
c. Most of the repressor protein is bound to tryptophan, forming the tryptophan-repressor protein complex.
d. RNA polymerase can bind to the operator.
e. Enzymes for tryptophan synthesis are not being manufactured.

RELATED ACTIVITIES

• Although you may conceptually understand the process of DNA replication, there is no substitute for actually working with physical models and demonstrating mechanically how new DNA can be built from an existing template. Using paper shapes, create 20 models of each of the four DNA bases. (The exact shape doesn't matter, as long as the four bases are different from one another. Making photocopies of the shapes will save you some work!) Construct a molecule of DNA that is 10 base pairs long. Figure out how this molecule can serve as a template for the formation of two identical molecules of DNA. (Hint: You might use the DNA segment shown in Figure 14.4 as a starting point.) Once you have mastered basic DNA replication, practice introducing mutations to see how these changes affect the replication of new DNA molecules.

- Consider the discussion of radiation poisoning experienced by individuals who survived the atomic blasts at Hiroshima and Nagasaki. Using the Internet, search for other locations where individuals were exposed to excess radiation. Compose a one-page summary describing the effects of radiation exposure in humans and the amount of time it takes for a population to recover from such exposure. In your summary, indicate how exposure to radiation affects DNA and cells in general.

- Until very recently, the history of science was full of examples of intelligent and capable women who were either prevented from becoming scientists or, if they did manage to do research, had their work ignored simply because they were female. Rosalind Franklin, discussed in this chapter, is only one such example of this unfortunate gender bias in science. Consult an encyclopedia or other reference source and write a one-page essay about another woman scientist who struggled to have her work properly recognized in the male-dominated world of research.

ANSWERS AND EXPLANATIONS

Factual Knowledge

1. e. The work of geneticists before 1928 had already provided this information and set the stage for the molecular study of genetic inheritance that ultimately led to the discovery of the structure of DNA. For more information, see Section 14.1, An Overview of DNA and Genes.

2. False. Before the work of Franklin, Wilkins, Watson, and Crick, the structure of DNA was incorrectly thought to be relatively simple. Geneticists did not believe it could provide the necessary complexity to be the hereditary material. Because of its chemical diversity, protein was initially believed to be the best candidate. For more information, see Section 14.1, An Overview of DNA and Genes.

3. False. Although each separate DNA strand is held together by strong covalent bonds, the two complementary strands are kept in close proximity to one another with relatively weak hydrogen bonds. Refer to Figure 14.3 for additional information. See also Section 14.2, *DNA is built from two helically wound polynucleotides.*

4. c. Hydrogen bonds connect the bases on opposite strands of the DNA molecule, which is the reason the double helix can be unwound and separated relatively easily. All other bonds in the molecule are covalent. Refer to Figure 14.3 for additional information. See also Section 14.2, *DNA is built from two helically wound polynucleotides.*

5. c. The DNA double helix is structurally very similar to a spiral staircase, with the hydrogen bonds between bases representing the steps and the railings corresponding to the opposing strands of nucleotides. For more information, see Section 14.2, *DNA is built from two helically wound polynucleotides.*

6. a. The fundamental factor that makes one individual different from another is the specific sequence of DNA bases in our genes. For more information, see Section 14.2, *DNA's structure explains its function.*

7. e. All of these will be true if the DNA molecule replicates correctly. For more information, see Section 14.3, How DNA Is Replicated.

8. d. Even though DNA replication typically occurs millions of times during the life of a multicellular organism, it is a remarkably accurate process. Those errors that do occur are usually corrected with a high degree of reliability. For more information, see Section 14.4, *Few mistakes are made in DNA replication.*

9. a. The amount of DNA packed within a human cell is approximately a little more than 2 meters in length. If all of the DNA in a human body were extended end to end, the distance would roughly equal 130 times the distance from Earth to the sun. For more information, see Section 14.6, DNA Packing in Eukaryotes.

10. e. Liver cells are representative of your body cells in general, and therefore all of these DNA features would be expected to exist. For more information, see Section 14.7, *Different cells in eukaryotes express different genes.*

11. c. Recall that DNA in eukaryotes is highly compacted. This is accomplished through the formation of histone spools in which DNA is wound around a complex of histone proteins. Review Figure 14.10 for more information. See also Section 14.6, DNA Packing in Eukaryotes.

12. True. Most prokaryotes are not long-lived, and so short-term responses to environmental changes, such as nutrient supply, are the main focus of their gene activity. Review Figure 14.11 for an example of bacterial gene regulation in response to the environment. See also Section 14.7, *Organisms can turn genes on or off in response to environmental cues.*

13. a. Transposons are DNA sequences that are capable of changing position within a chromosome. The hemoglobin, insulin, and crystallin protein genes are specialty genes that are expressed only in cells that need those products. Ribosomal RNA genes are needed in all cells that carry out protein synthesis, and so qualify as housekeeping genes. For more information, see Section 14.7, *Different cells in eukaryotes express different genes.*

14. d. All of these are ways in which gene expression can be regulated, but transcriptional control is clearly the

most common mechanism. For more information, see Section 14.8, *Most genes are controlled at the transcriptional level.*

15. b. DNA bases
    d. DNA replication
    a. mutation
    c. operator
    e. genome

For more information, see Section 14.1, An Overview of DNA and Genes; Section 14.2, *DNA is built from two helically wound polynucleotides*; Section 14.3, How DNA Is Replicated; and Section 14.4, *Few mistakes are made in DNA replication.*

Conceptual Understanding

1. d. Recall that since A pairs with T, there should be equal numbers of both bases in the molecule, totaling 44 percent. Therefore, the remaining 56 percent must be made from equal amounts of C and G, meaning we would expect half this amount (28 percent) to be the total percentage of cytosine present in the molecule. For more information, see Section 14.2, *DNA is built from two helically wound polynucleotides.*

2. b. Following the Watson-Crick base pairing rules (A-T; G-C), we can easily determine the sequence of the complementary strand. For more information, see Section 14.2, *DNA is built from two helically wound polynucleotides.*

3. e. Hydrogen bonds between complementary bases on opposite strands hold the DNA molecule together. An enzyme that separates the strands must therefore break these bonds. For more information, see Section 14.2, *DNA is built from two helically wound polynucleotides.*

4. b. In DNA replication, each strand of the original molecule serves as a template for the formation of a new strand. Thus, both new molecules would have half of its nucleotides (and the chemical label) from the original molecule. For more information, see Section 14.3, How DNA Is Replicated.

5. a. The introduction of a mismatch error has the potential of changing a particular allele permanently. In some instances, this change can result in a defective gene that may then serve as the basis of a genetic disorder. For more information, see Section 14.4, *Few mistakes are made in DNA replication.*

6. e. DNA repair is a multistep process involving all of the events and structures listed. For more information, see Section 14.4, *Normal gene function depends on DNA repair.*

7. c. In XP, a recessive allele codes for the production of a faulty protein in the normal DNA repair mechanism. Thus, when ultraviolet light damages DNA in skin cells, the damage cannot be repaired and the physical symptoms of XP, such as severe blisters and skin cancer, emerge. For more information, see Section 14.4, *Normal gene function depends on DNA repair.*

8. c. Keep in mind that for an allele to be passed on to the next generation, it must be present in either the sperm or the egg cells contributed by the parent. If the mutation occurs elsewhere in the body, it will have no chance of being inherited by subsequent generations. For more information, see Section 14.4, *Few mistakes are made in DNA replication.*

9. c. Such tightly packed DNA will not undergo transcription because the necessary enzymes that control transcription cannot gain access to the DNA. For more information, see Section 14.6, DNA Packing in Eukaryotes.

10. d. These three clues are all characteristics of eukaryotic DNA, but we have no information that permits us to decide from which kingdom of eukaryotes the sample came. For more information, see Section 14.5, Genome Organization.

11. e. Recall that the tryptophan operator (see Figure 14.13) is a method of transcriptional control used in prokaryotes. For more information, see Section 14.8, *Most genes are controlled at the transcriptional level.*

12. c. Recall that rRNA genes are considered to be housekeeping genes; therefore, they should be expressed in all cell types. The product insulin is differentially expressed in the cells of the pancreas. For more information, see Section 14.7, *Different cells in eukaryotes express different genes.*

13. d. Because tryptophan is lacking, it would not be present in quantities sufficient to form the active tryptophan-repressor protein complex. RNA polymerase is thus not prevented from binding to the operator. This in turn causes the transcription of mRNA that codes for the production of the enzymes used in tryptophan synthesis. For more information, see Section 14.8, *Most genes are controlled at the transcriptional level.*

CHAPTER 15 | From Gene to Protein

GETTING STARTED

Below are a few questions to consider before reading Chapter 15. These questions will help guide your exploration and assist you in identifying some of the key concepts presented in this chapter.

1. What did British physician Archibald Garrod believe was the cause of inherited human metabolic disorders such as alkaptonuria?

2. What are the functional differences between the three types of RNA—messenger RNA (mRNA), ribosomal RNA (rRNA), and transfer RNA (tRNA)?

3. What is a codon?

4. What is the primary difference between an intron and an exon?

5. In the genetic code, what sequence constitutes the universal start codon?

6. How are frameshift mutations similar to losing your place on a multiple-choice answer sheet?

7. During protein translation, what is wobble?

8. Which genetic disease results in the formation of curved and distorted red blood cells?

9. How were scientists able to confirm the identity of the Iceman using DNA technology?

A GUIDE TO THE READING

The following concepts typically give students the most difficulty when exploring the content in Chapter 15 for the first time. For each concept, one or more references have been identified that may help you gain a better understanding of these potentially problematic areas.

Three Types of RNA

The cell makes use of three different types of RNA, each with a specific function associated with the production of proteins from genes. Messenger RNA represents a copy of the sequence of nucleotides present in a gene. This information is then decoded to produce proteins by structures called "ribosomes." A major component of ribosomes is ribosomal RNA. The building blocks of proteins are amino acids. Amino acids are transferred to the ribosome on transfer RNA molecules for use in protein synthesis. Each of the three types of RNA molecules has a very specific role in the process of gene expression.

For more information on this concept, be sure to focus on

- Section 15.1, *Three types of RNA assist in the manufacture of proteins*
- Table 15.1, RNA Molecules and Their Functions

Transcription

The first step in the process of gene expression is the "copying" of the genetic information stored in the DNA molecule into a form that may be used in the cytoplasm. This copy is made using RNA, resulting in the production of an mRNA molecule. It is important to note that only one of the two complementary strands of the DNA molecule contains the proper sequence used for the production of the mRNA molecule. The strand that contains the information is referred to as the "template strand." The starting point for a gene is marked by a specific sequence called the "promoter," so named because it "promotes" the attachment of the RNA polymerase molecule responsible for producing the mRNA molecule. It is also important to remember that during transcription, the ribonucleotide uracil is substituted for the deoxynucleotide thymine. RNA polymerase continues to produce the mRNA molecule, following the sequence present in the gene, until it reaches the "terminator"—a specific sequence that signals the completion of the transcription process. In eukaryotes, this process results in an mRNA molecule composed of both introns and exons. The introns contain extraneous information and must be removed before translation through the process of RNA processing, which includes RNA splicing. This completes the formation of a complete mRNA molecule.

For more information on this concept, be sure to focus on

- Section 15.3, *Transcription: Information Flow from DNA to RNA*
- Figure 15.3, RNA Polymerase Transcribes DNA-Based Information into RNA-Based Information
- Figure 15.4, In Eukaryotes, Introns Must Be Removed Before an mRNA Leaves the Nucleus

Translation

The second step in the process of gene expression is the decoding of the genetic information copied from the genes into its final form, protein. Information decoding is performed by specialized structures present in the cytoplasm, called "ribosomes." Ribosomes attach to the mRNA molecule (produced through transcription) and begin the process of translating the sequence of nucleotides, using the genetic code (Figure 15.5 in your textbook), into a sequence of amino acids. This process results in the formation of an intact protein. In addition to mRNA, ribosomes must also interact with tRNA molecules that transport amino acids. Transfer RNA molecules provide the key to translating the genetic code. These molecules contain a three-nucleotide sequence (called the "anticodon") that is complementary to the codon sequences present in the mRNA molecule. The codon and anticodon match precisely, using Watson-Crick base pairing rules (Chapter 14). Because there are 64 possible combinations of codons, this means there are a total of 64 different tRNA molecules, each with a complementary anticodon. Each of these different tRNA molecules, however, carries only one of the 20 different amino acids. Therefore, it is the process of tRNA production that matches anticodon sequences to the appropriate amino acid molecule, ensuring the accuracy of the translation process.

For more information on this concept, be sure to focus on

- Section 15.4, The Genetic Code
- Section 15.5, Translation: Information Flow from mRNA to Protein
- Figure 15.5, The Genetic Code
- Figure 15.7, Transfer RNA Delivers Amino Acids Specified by mRNA Codons

- Figure 15.8, In Translation, Information Coded in mRNA Directs the Synthesis of a Protein with a Precise Amino Acid Sequence

Mutations

As we saw in the current and previous chapters, mutations alter a gene sequence, and this may result in the formation of a new form of the gene (allele). The chapter mentions the three main types of mutations that may occur. Substitution mutations are caused by the substitution of one nucleotide with another (such as the substitution of a T for a C). This may occur as an error made during DNA replication. Not all substitution mutations are significant in terms of gene expression. For such a mutation to affect a gene product, the error must occur not only within a gene but also within a coding region (exon) of the gene. If the error occurs within an intron region, this segment will be removed before translation and therefore will have no impact on the protein product. In some cases, a substitution mutation may also be "silent." Silent mutations are possible because of the redundancy of the genetic code (more than one codon may specify the same amino acid). For example, the codons GAA and GAG both code for the amino acid glutamate. As a result, if a substitution mutation were to occur in the third position (A for G or G for A), the amino acid would remain unchanged in the final protein product. Insertion and deletion mutations are potentially more disruptive. In these cases, nucleotides are inserted or deleted into the gene sequence. This may result in a genetic "frameshift," where the codons become shifted by one base, disrupting the meaning of the information, much like accidentally skipping a question on an answer sheet for a multiple-choice exam.

For more information on this concept, be sure to focus on

- Section 15.4, The Genetic Code
- Section 15.6, The Effect of Mutations on Protein Synthesis
- Figure 15.5, The Genetic Code
- Figure 15.9, A Change in the DNA Sequence Translates into a Change in the Amino Acid Sequence of the Protein

TYING IT ALL TOGETHER

Several concepts presented in this chapter build on those presented in previous chapters and may also be revisited and discussed in greater detail in subsequent chapters, including

Protein Synthesis

- Chapter 5—Section 5.8, *Proteins are built from amino acids*

Gene Mutation

• Chapter 12—Section 12.1, *Gene mutations are the source of new alleles*

Chromosomal Basis of Inheritance

• Chapter 13—Section 13.1, The Role of Chromosomes in Inheritance

Genetic Disorders

• Chapter 13—Section 13.4, Human Genetic Disorders

DNA Replication

• Chapter 14—Section 14.3, How DNA Is Replicated

Control of Gene Expression

• Chapter 14—Section 14.8, How Cells Control Gene Expression

PRACTICE QUESTIONS

Factual Knowledge

1. For the information contained in a gene to be used to produce a functioning protein,
   a. DNA must be replicated.
   b. information must be transcribed into mRNA and then translated into amino acids.
   c. tRNA must be transcribed into rRNA and then translated into amino acids.
   d. ribosomes must be converted from rRNA into mRNA.
   e. the mRNA message must be sent from the cytoplasm to the cell nucleus.

2. Genes code for which of the following biological molecules?
   a. enzymes
   b. structural proteins
   c. transport proteins
   d. defense proteins
   e. all of the above

3. A sequence of DNA that contains information for the synthesis of RNA molecules used in the manufacture of proteins is also known as a(n)
   a. intron.
   b. mutation.
   c. gene.
   d. polymerase.
   e. codon.

4. The process of gene transcription begins with the
   a. binding of RNA polymerase to a region of DNA called the "promoter."
   b. removal of introns from the newly formed mRNA.
   c. joining of rRNA with various ribosomal proteins.
   d. attachment of an mRNA molecule to the ribosome.
   e. insertion of new DNA bases into the gene.

5. Which of the following events does *not* occur during the process of translation?
   a. The information in mRNA is used to guide the assembly of amino acids into proteins.
   b. (C)ytosine in DNA pairs with (G)uanine in RNA, and (T)hymine in DNA pairs with (A)denine in RNA.
   c. Start and stop codons present in the mRNA molecule help control the process.
   d. Amino acids are attached to the end of the new protein molecule under construction.
   e. The ribosome interacts with mRNA in the cytoplasm.

6. During protein translation, "wobble" refers to
   a. the ability of a tRNA molecule to bind to more than a single unique codon.
   b. the ability of a cell to utilize either introns or exons as protein coding information.
   c. the redundancy of the genetic code.
   d. the movement of ribosomes along an mRNA molecule.
   e. none of the above

7. There are two binding sites on a molecule of tRNA—one for the attachment of an amino acid, and the other where the tRNA anticodon binds to the corresponding codon on the mRNA molecule. (True or False)

8. Refer to Figure 15.8. The small red triangle represents
   a. the last amino acid added to the new protein.
   b. the first amino acid added to the new protein.
   c. the anticodon of the last tRNA used to attach a new amino acid.
   d. RNA polymerase.
   e. a codon.

9. In a ribosome in the process of translating a molecule of mRNA, a maximum of _____ codon(s) can be occupied by tRNA at any one time.
   a. 1
   b. 2
   c. 3
   d. 64
   e. depends on the kind of mRNA

10. Consider the DNA template shown in the middle of Figure 15.9. If the C at the start of the sequence of bases is deleted, this genetic mutation will result in

a. a frameshift that affects the entire mRNA molecule.

b. a change in the first amino acid coded for, but not in any others.

c. a change in the last amino acid coded for, but not in any others.

d. all the amino acids coded for now being the same.

e. no effect on either the mRNA or the protein manufactured.

11. Similar to genes that code for proteins, the products of genes that code for rRNA and tRNA must also undergo translation. (True or False)

12. Match each term with the best description.
___ codon
___ mRNA
___ transcription
___ translation
___ intron

a. conversion of mRNA information to amino acid sequence

b. three-base combination of nucleotides in mRNA

c. a process occurring within the cell nucleus

d. unused portion of an mRNA transcript

e. specifies sequence of amino acids in proteins

Conceptual Understanding

1. A murder has occurred, and you are asked to help solve it. The police bring you a sample from the crime scene of what they believe is the killer's DNA and ask you for a chemical analysis. Your study of this sample reveals the presence of adenine, thymine, ribose, and uracil, leading you to conclude that the sample is

a. pure DNA.

b. pure RNA.

c. probably a mixture of DNA and RNA.

d. probably a mixture of rRNA and mRNA.

e. definitely not from the killer.

2. The work of Archibald Garrod and William Bateson on inherited human metabolic disorders provided some of the first clues that genes work by controlling the production of proteins. (True or False)

3. If you expose a cell to chemicals that specifically disrupt the function of RNA polymerase, which of the following processes will be most directly affected?

a. transcription

b. translation

c. DNA replication

d. rate of mutation

e. evolution of the genetic code

4. Refer to the information on the genetic code provided in Figure 15.5. Use this information to determine how many amino acids are coded for by the mRNA sequence AUGCGCAGUCGGUAG.

a. 0

b. 4

c. 5

d. 15

e. cannot determine from the information given

5. A small segment of DNA on the template strand contains the base sequence CGT. If an mRNA transcript is made that includes this sequence, what would be the anticodon on the tRNA that would bind to this corresponding mRNA sequence?

a. CGT

b. GCA

c. CGU

d. GCT

e. cannot determine from the information given

6. Without the _____ site on a molecule of DNA, _____ cannot bind to the DNA to begin the process of _____.

a. active; introns; tRNA translation

b. promoter; RNA polymerase; mRNA transcription

c. promoter; rRNA; ribosome activation

d. attachment; ribosomal proteins; frameshift mutation

e. stop codon; amino acids; translation

7. Consider a complete, functional piece of mRNA that contains a total of 66 codons. What is the maximum number of amino acids that could be present in the protein coded for by this mRNA?

a. 21

b. 22

c. 64

d. 65

e. 66

8. A triplet base sequence on the template strand of DNA reads ATT. What will be the corresponding mRNA codon, tRNA anticodon, and amino acid coded for by this DNA?

a. TAA; UTT; methionine

b. TAA; AUU; no amino acid (= stop codon)

c. UAA; AUU; no amino acid (= stop codon)

d. CGG; GCC; alanine

e. CGG; ATT; isoleucine

9. Refer to Figure 15.10. If the original triplet DNA codon CTC had mutated into CTT instead of CAC as shown, what would have been the consequence for the hemoglobin formed from this alternative mutation?

a. Normal hemoglobin would have been made.

b. Sickle-cell hemoglobin would have been made.

c. Threonine would have been substituted for glutamate instead of valine.

d. Methionine would have been substituted for glutamate instead of valine.

e. none of the above

10. Before mutation, a sequence of DNA reads GAGCCTATGCCAGTA. After the mutation, the sequence reads GAGCGTACGCCATTA. Which of the following best explains the change in DNA that has occurred?

a. There was a single base deletion.

b. There was a single base substitution.

c. There were multiple base deletions.

d. There were multiple base substitutions.

e. There were multiple base insertions.

11. Consider the chapter's Biology Matters box, "One Allele Makes You Strong, Another Helps You Endure." Which of the following human genotypes would the genetically engineered knockout mice most closely resemble?

a. *RR*

b. *RX*

c. *XX*

d. *XR*

e. none of the above

RELATED ACTIVITIES

• Cut out 24 half-inch squares of paper and label six of them A, six T, six C, and six G. These represent DNA bases. Cut out another 24 squares and label six more A, six U, six C, and six G. These represent RNA bases. Now randomly arrange the DNA bases into a linear sequence. What is the corresponding mRNA that would be manufactured during transcription of this DNA? Once you have completed this task, rearrange the DNA and make new mRNA as many times as necessary until you are convinced you could teach someone how transcription operates within the cell nucleus.

• After you have completed the activity above, make a second set of RNA bases. These will represent tRNA. Again, make a sequence of DNA bases and the corresponding mRNA. Now indicate the tRNA anticodons that would be needed to match the mRNA transcript. Refer to the genetic code in Figure 15.5 and write down the amino acid sequence that this segment of DNA codes for. Repeat the process until you are comfortable with converting DNA and mRNA information into an amino acid sequence, as occurs during the process of translation.

• Consider the chapter's Biology in the News box, "Atrazine Hurts Animals." Using the Internet or other resources, learn about the usage of atrazine in your area or another specific geographic region that interests you.

Where and how is it being used? What are the levels in the local water? Are there animal populations in the region that are at risk from this chemical usage? Write a one-page paper intended as a letter to the editor of a local paper to educate the public about this issue.

ANSWERS AND EXPLANATIONS

Factual Knowledge

1. b. Protein synthesis is a two-step process involving transcription in the cell nucleus followed by translation in the cytoplasm. All the other choices are either factually inaccurate or, if accurate, need not necessarily take place for protein synthesis to occur. For more information, see Section 15.1, How Genes Work.

2. e. All of the molecules listed are types of proteins, which are coded for by genes. For more information, see Section 15.1, *Genes contain information for building RNA molecules.*

3. c. This is the functional definition of one certain type of gene. For more information, see Section 15.1, *Genes contain information for building RNA molecules.*

4. a. The first step in the transcription process is the binding of RNA polymerase, the primary enzyme responsible for the process, to a specific sequence that signals the start of the gene called the "promoter." For more information, see Section 15.3, Transcription: Information Flow from DNA to RNA.

5. b. The pairing of DNA nucleotides with RNA nucleotides occurs during transcription, not translation. For more information, see Section 15.5, Translation: Information Flow from mRNA to Protein.

6. a. Some tRNA molecules are capable of recognizing more than a single codon. This is due to the fact that the third position of anticodons can pair with different bases, resulting in "wobble." This is the reason why most cells contain only about 40 different tRNA molecules instead of the expected 60-plus tRNA molecules that would be needed to account for each of the 64 codons (Figure 15.5). For more information, see Section 15.5, Translation: Information Flow from mRNA to Protein.

7. True. By having these two binding sites, tRNA can interact with both a specific amino acid and a unique codon on a strand of mRNA that is being read by a ribosome. For more information, see Section 15.2, How Genes Guide the Manufacture of Proteins, and Section15.5, Translation: Information Flow from mRNA to Protein.

8. b. The triangle represents the amino acid methionine, carried by the tRNA that binds to the start codon. For

more information, see Section 15.5, Translation: Information Flow from mRNA to Protein.

9. b. Because there are two attachment sites on the ribosome, two mRNA codons (and therefore two tRNA anticodons) may be present at any one time. Review Figure 15.8 to visualize the spatial configuration of the mRNA-ribosome interaction. For more information, see Section 15.5, Translation: Information Flow from mRNA to Protein.

10. a. Deletion mutations of the sort described here typically result in a frameshift that ripples through the remainder of the base sequences. The result is usually an incomplete or malfunctioning protein. For more information, see Section 15.6, *Mutations can alter one or many bases in a gene's DNA sequence.*

11. False. In genes that code for rRNA and tRNA, the RNA is actually the end product and therefore does not undergo translation into a protein. Only mRNA gets translated. For more information, see Section 15.2, How Genes Guide the Manufacture of Proteins.

12. b. codon
    e. mRNA
    c. transcription
    a. translation
    d. intron
    For more information, see Section 15.2, How Genes Guide the Manufacture of Proteins; Section 15.3, Transcription: Information Flow from DNA to RNA; and Section 15.4, The Genetic Code.

Conceptual Understanding

1. c. The available evidence indicates that both DNA and RNA are present in the sample. The logic here is that thymine is unique to DNA, and both ribose and uracil are found only in RNA. With this evidence, however, you can say nothing about the types of RNA present in the sample, or whether the sample is actually from the killer. For more information, see Section 15.1, *Genes contain information for building RNA molecules.*

2. True. Both of these researchers made the link between genes and enzymes (types of proteins) through studies of inherited metabolic disorders in humans. For more information, see Section 15.1, How Genes Work.

3. a. RNA polymerase is the enzyme that binds to the promoter site on DNA to begin the process of gene transcription. For more information, see Section 15.3, Transcription: Information Flow from DNA to RNA.

4. b. The first four sets of triplet bases code for the amino acids methionine (start codon), arginine, serine, and arginine, respectively. The last codon is a ter-

mination (stop) signal. Thus, the polypeptide coded for by this segment of mRNA will have a total of four amino acids. For more information, see Section 15.4, The Genetic Code.

5. c. A DNA triplet of CGT in the template strand would yield an mRNA codon of GCA. Therefore, the complementary tRNA anticodon would be CGU. For more information, see Section 15.5, Translation: Information Flow from mRNA to Protein.

6. b. This critical step required to initiate transcription is depicted in Figure 15.3. For more information, see Section 15.3, Transcription: Information Flow from DNA to RNA.

7. d. A functional strand of mRNA must have both a start and a stop codon. The start codon typically codes for the amino acid methionine, which may or may not end up being a part of the final protein. However, the stop codon would not code for an amino acid. Thus, with 66 codons in the mRNA, there could be as many as 65 amino acids in the final protein product. For more information, see Section 15.4, The Genetic Code.

8. c. The complementary base-pairing rules of DNA-to-RNA and RNA-to-RNA give the codon and anticodon. Recall also that there is no thymine in RNA; uracil and adenine are complementary bases for RNA. Figure 15.5 indicates that the mRNA codon UAA is a stop signal and does not code for any amino acid. For more information, see Section 15.4, The Genetic Code.

9. a. If the alternative mutation had occurred, the new mRNA codon would be GAA. Using Figure 15.5, we see that GAA also codes for glutamate, just like the original DNA message, and therefore normal hemoglobin would have been produced. For more information, see Section 15.4, The Genetic Code.

10. d. Because the total number of bases is the same, it is unlikely that either deletions or insertions occurred. In this case, however, simply comparing each strand base for base shows that the more likely explanation is that multiple substitutions took place along the sequence. For more information, see Section 15.6, *Mutations can alter one or many bases in a gene's DNA sequence.*

11. c. Recall from the explanation of the article that in humans, the *XX* genotype confers two truncated versions of the *ACTN3* gene, which is important for the function of fast-twitch muscle fibers. In the knockout mice, the *ACTN3* gene was removed, essentially mimicking the human *XX* genotype, which is found to be more common in high-endurance athletes. For more information, see the Biology Matters box, "One Allele Makes You Strong, Another Helps You Endure."

CHAPTER 16 | DNA Technology

GETTING STARTED

Below are a few questions to consider before reading Chapter 16. These questions will help guide your exploration and assist you in identifying some of the key concepts presented in this chapter.

1. How was Edunia, the "plantimal," created? What significance does this flower have beyond its artistic merit?

2. What is the natural function of restriction enzymes?

3. Why must DNA be heated or treated with chemicals during the course of a DNA hybridization experiment?

4. What would you expect to find in a DNA library?

5. What are the advantages of using the polymerase chain reaction (PCR) for gene cloning?

6. Why would DNA fingerprinting not be useful to distinguish between identical twins?

7. What percentage of the world's soybeans, corn, cotton, and canola crops is genetically modified?

8. How did gene therapy help Ashanthi DeSilva lead a normal life?

A GUIDE TO THE READING

The following concepts typically give students the most difficulty when exploring the content in Chapter 16 for the first time. For each concept, one or more references have been identified that may help you gain a better understanding of these potentially problematic areas.

Restriction Enzymes and Gel Electrophoresis

Restriction enzymes are capable of breaking the covalent bonds that hold the long, individual strands of DNA together. What makes restriction enzymes useful for the analysis of DNA is the specificity they display in selecting a region in which to cut the DNA molecule. Each restriction enzyme has a target sequence, a specific sequence of nucleotides that is typically 10 base pairs or fewer in length, which must be present for the enzyme to be able to attach to the DNA molecule and cut the two strands. This target sequence varies from enzyme to enzyme (see Figure 16.9 in your textbook). Because identical copies of a DNA molecule would have the same set of target sequences, a restriction enzyme would cut these copies in the exact same location. However, if the copies of DNA under analysis were different in their nucleotide sequences, a single restriction enzyme would recognize and cut only those DNA molecules with the correct target sequence. This would result in DNA fragments of differing sizes. In a test tube, it would be impossible to visualize these DNA fragments.

Therefore, scientists have devised a way to separate these DNA fragments on the basis of size using gel electrophoresis. In gel electrophoresis, the DNA samples are loaded onto a slab of a Jell-o–like matrix and subjected to an electrical charge. The DNA fragments, which are negatively charged, are attracted to the positively charged end of the gel slab and slowly migrate in that direction (see Figure 16.10). Larger fragments move more slowly because they have a more difficult time making their way through the matrix, whereas smaller fragments move more rapidly. Fragments of the same size tend to move as a group, accumulating in a region called a "band." By analyzing these separated bands, scientists can determine whether a sample of DNA contained one or more target sequences for a particular restriction enzyme. This technique is particularly useful as a DNA fingerprinting tool.

For more information on this concept, be sure to focus on

- Section 16.7, *Enzymes are used to cut and join DNA*
- Section 16.7, *Gel electrophoresis sorts DNA fragments by size*
- Figure 16.9, Restriction Enzymes Cut DNA at Specific Places
- Figure 16.10, DNA Fragments Can Be Separated by Gel Electrophoresis
- Figure 16.11, Restriction Enzymes and Gel Electrophoresis Can Be Used to Identify the Sickle-Cell Allele

Polymerase Chain Reaction

PCR is a technique that uses the enzyme DNA polymerase in a test tube to produce millions of copies of a piece of DNA in a short time. The key to understanding PCR is realizing that a portion of the sequence of DNA a scientist wishes to copy (or amplify) must already be known. This is because the DNA polymerase enzyme is capable only of extending a DNA strand; it cannot build a DNA strand from scratch. Therefore, PCR requires the use of primers (see Figure 16.13), which are short segments of DNA that match the two ends of the sequence of DNA targeted for copying. In a typical PCR reaction, the two strands of template DNA are separated using heat. By cooling the reaction a little, the primer DNA strands are allowed to bind to their complementary sequences located on the target strands. DNA polymerase can then begin to extend the primer using the original DNA as a template. This results in two identical copies of the target DNA. This cycle is repeated 20 to 30 times, each time doubling the number of copies of the target DNA in the test tube, until billions of copies of the DNA sequence of interest are obtained.

For more information on this concept, be sure to focus on

- Section 16.7, PCR is used to amplify small quantities of target DNA
- Figure 16.13, Small Amounts of DNA Can Be Amplified More than a Millionfold through PCR

Cloning

As we saw in the chapter, cloning can mean several things. In DNA cloning, a single copy of a gene is isolated and inserted into some type of vector for easy copying and subsequent transfer to another organism. On a larger scale, the cloning of entire organisms is referred to as "reproductive cloning," whereas the cloning of tissues or organs is termed "therapeutic cloning." If you are familiar with the story of Dolly the sheep, you are aware of the process of reproductive cloning, in which an offspring that is an exact genetic duplicate of an

existing individual is produced. The process involves the transfer of a diploid nucleus isolated from the cell of an individual into an egg cell from which the nucleus has been removed. In this case, the imported diploid nucleus provides the entire genetic blueprint for the development of the egg into an embryo, a fetus, and ultimately a newborn. Because the DNA present in the newborn comes entirely from an existing individual, the newborn is genetically identical to the DNA donor and therefore represents a clone of this individual. In therapeutic cloning, stem cells are harvested from the developing embryo. In this case, these genetically identical stem cells may then be implanted or transferred to the DNA donor to treat a particular disorder, such as a degenerative neurological disease, or to restore the normal function of a tissue. The key to both these types of cloning procedures is the fact that the resulting stem cells or organisms are all genetic duplicates of the original donor.

For more information on this concept, be sure to focus on

- Section 16.3, Reproductive Cloning of Animals
- Section 16.7, DNA cloning is a means of propagating recombinant DNA
- Figure 16.5, Cloning Sheep: How Dolly Came to Be

TYING IT ALL TOGETHER

Several concepts presented in this chapter build on those presented in previous chapters and may also be revisited and discussed in greater detail in subsequent chapters, including

DNA Structure and Replication

- Chapter 14—Section 14.3, How DNA Is Replicated

Single Gene Mutations and Disease

- Chapter 15—Section 15.6, The Effect of Mutations on Protein Synthesis

The Immune System

- Chapter 32—Section 32.4, Third Line of Defense: The Adaptive Immune System

Stem Cells and Development

- Chapter 33—Section 33.3, Human Reproduction: From Fertilization to Birth

PRACTICE QUESTIONS

Factual Knowledge

1. Which of the following are products of genetic engineering?
   a. establishment of dog breeds
   b. agricultural crops with enhanced nutritional value
   c. DNA fingerprinting
   d. Restriction Fragment Length Polymorphisms (RFLPs)
   e. all of the above

2. The primary reason why the same basic techniques can be used to analyze the DNA from species as diverse as bacteria and humans is that
   a. all cells are identical.
   b. every organism has the same amount of DNA.
   c. the DNA sequences of all organisms are the same.
   d. DNA has a consistent structure in all organisms.
   e. all of the above

3. To study the genetic basis of the inherited human disease sickle-cell anemia, a biologist first isolates DNA from an affected individual's cells. Which of the following would be the next step in this process?
   a. Use restriction enzymes to break the DNA into smaller fragments.
   b. Apply electrical current across the gel to separate DNA fragments.
   c. Visualize DNA fragments in the gel using a stain.
   d. Load DNA fragments onto a gel.
   e. Use PCR to produce many copies of the target DNA sequence.

4. Which of the following statements about restriction enzymes is *false*?
   a. They work on DNA from all types of organisms.
   b. They are used to glue together short segments of DNA.
   c. They come in many varieties, each with its own DNA target sequence.
   d. They are highly specific for their DNA target sequences.
   e. They are believed to be a primitive type of immune system used by bacteria to degrade viral DNA.

5. DNA probes are
   a. large segments of double-stranded DNA.
   b. enzymes used in PCR.
   c. short, single-stranded segments of DNA used in DNA hybridization experiments.
   d. used to generate restriction fragment length polymorphisms.
   e. none of the above

6. In gel electrophoresis, the distance traveled by a DNA fragment
   a. corresponds to the size of the fragment.
   b. depends on which ligase was used to make the fragments.
   c. is related to the species of organism from which the DNA came.
   d. depends on whether the fragment is attached to a plasmid.
   e. all of the above

7. Which of the following is a critically important tool used in experiments involving DNA hybridization?
   a. restriction enzymes
   b. DNA probes
   c. heat
   d. radioactive or chemical labels
   e. all of the above

8. A gene is said to be cloned if
   a. the DNA sequence of the gene is known.
   b. the function of the gene is known.
   c. there is a DNA probe for the gene.
   d. the gene has been isolated and copied.
   e. it has been used by scientists for gene therapy.

9. Reproductive cloning, the creation of genetically identical animals, is an exciting technology that can provide important insights into basic biology but lacks practical applications. (True or False)

10. Which of the following is *not* necessary to execute a successful PCR?
    a. all four DNA bases
    b. short single-stranded DNA primers
    c. DNA polymerase
    d. a restriction enzyme
    e. a target DNA sequence

11. In genetic engineering, genes can be inserted from one organism into another or back into the original organism using which of the following techniques?
    a. PCR
    b. gene gun
    c. DNA hybridization
    d. gel electrophoresis
    e. gene therapy

12. Which of the following DNA sources would be suitable for DNA fingerprinting analysis?
    a. blood
    b. skin
    c. semen
    d. bone
    e. all of the above

13. Which of the following statements regarding DNA technology is *false*?
    a. DNA technology can be used to alter the performance of a modified organism.
    b. DNA technology can be used to produce multiple copies of a desired DNA sequence.
    c. DNA technology can be used to generate many copies of specific genes.
    d. DNA technology can enhance the production of a specific gene product.
    e. DNA technology can be used to remove undesired alleles from a multicellular organism.

14. In human gene therapy, doctors
    a. seek to correct genetic disorders by repairing the genes that cause them.
    b. treat the symptoms of a disorder, but not the underlying genetic cause.
    c. must work at the embryonic stage of development to be effective.
    d. know with certainty that the therapy alone improves the patient's condition.
    e. have been able to use the viral method to deliver modified genes.

15. Match each term with the best description.
    __ DNA probe
    __ clone
    __ plasmid
    __ recombinant DNA
    __ gene therapy
    a. vector for DNA transfer between cells
    b. effort to fix the genes that cause disorders
    c. DNA sequence that can pair with a particular gene
    d. DNA containing sequences from multiple sources
    e. isolated gene of which many copies are made

Conceptual Understanding

1. Which of the following would *not* be considered a concern for critics of genetic engineering?
    a. Genetically modified plants or animals could spread to wild species.
    b. Genetically modified foods may cause allergic reactions in unsuspecting consumers.
    c. Genetic engineering of herbicide-resistant plants may lead to increased use of herbicides.
    d. Genetic engineering of plants containing Bt toxin may be harmful to beneficial insect species.
    e. none of the above

2. Imagine that sometime in the future human space explorers discover life on another planet and that this life has DNA as its genetic material. Which of the following could be used to study the genetics of this extraterrestrial life?
    a. cloning
    b. gel electrophoresis
    c. PCR
    d. restriction enzymes
    e. all of the above

3. DNA ligases and polymerases are key components of DNA technology because they help break down the organismal genome into DNA fragments. (True or False)

4. Imagine a gel through which DNA fragments have moved in response to an applied electrical current. The band on this gel that is farthest from the top (that is, from the place where the DNA fragments were added to the "well") represents the
    a. shortest fragments of DNA.
    b. longest fragments of DNA.
    c. restriction enzyme used to cut the DNA into fragments.
    d. ligase used to bind the DNA fragments together.
    e. DNA polymerase used to make copies of the DNA fragments.

5. Refer to Figure 16.11. At the bottom of the diagram is the result of a gel electrophoresis run using the DNA fragments from a sickle-cell allele and a normal allele that were generated using the restriction enzyme *Dde*I. Which of the fragments (*x*, *y*, or *z*) shown is the largest?
    a. *x*
    b. *y*
    c. *z*
    d. They are all the same size.
    e. cannot determine from the information given

6. A sample of DNA from a person suspected of having sickle-cell anemia is subjected to DNA hybridization using two probes—one that binds to the normal allele and another that binds to the sickle-cell allele. If both probes bind to the DNA, this individual
    a. is homozygous for the sickle-cell gene.
    b. is heterozygous for the sickle-cell gene.
    c. is normal.
    d. has sickle-cell anemia.
    e. none of the above

7. DNA cloning is beneficial because a cloned gene can be
    a. sequenced using automated sequencing machines.
    b. transferred to other cells of the same organism.
    c. transferred to cells in other organisms.
    d. used in DNA hybridization experiments.
    e. all of the above

8. Consider Figure 16.3. What would be the effect of using one type of restriction enzyme to cut the DNA containing the gene of interest, and then a different one to cut the plasmid DNA in the first step?
   a. Incompatible DNA fragments would be produced.
   b. A different ligase would be needed to join the DNA fragments.
   c. A different kind of plasmid would be needed as a vector.
   d. Bacteria would not be able to take up the plasmids produced.
   e. The results would be the same as those pictured in the figure.

9. A biologist intends to use PCR to perform a task. The biologist probably is trying to
   a. discover new genes.
   b. clone a gene.
   c. cut DNA into many small fragments.
   d. isolate DNA from a living cell.
   e. separate DNA fragments from one another.

10. In a criminal trial in which DNA fingerprinting has been used, a genetic "match" between the suspect and evidence left at the crime scene provides
    a. definitive proof that the suspect is guilty.
    b. a high probability that the suspect is guilty.
    c. a low probability that the suspect is guilty.
    d. definitive proof that the suspect is innocent.
    e. no useful information.

11. Prenatal screening in humans, HIV testing in humans, and genetic engineering for protection against insect attack in plants all
    a. are types of gene therapy.
    b. are examples of the application of DNA technology.
    c. require use of the same restriction enzyme.
    d. are carried out using gene guns.
    e. involve no ethical dilemmas or environmental risks.

12. Genetically engineered bovine growth hormone (BGH), which is highly effective for improving overall growth and milk production in cattle, remains a hotly debated issue because
    a. BGH is clearly hazardous to human health.
    b. BGH is an environmental hazard.
    c. BGH could drive traditional family farmers out of business.
    d. scientists remain unconvinced that BGH really works.
    e. all of the above

13. Gene therapy has been tested on a large number of patients with a wide variety of inherited genetic disorders, and in numerous cases it has produced a complete cure. (True or False)

14. On the basis of the chapter's Biology Matters box, "Have You Had Your GMO Today?," which of the following statements is *false*?
    a. Consumers will have the final say regarding the success of genetically modified crops.
    b. Genetically modified crops are used only for the production of breads and cereals.
    c. Genetically modified crops have allowed the United States to remain among the most agriculturally productive nations.
    d. The United States is one of the global leaders in the production of genetically modified crops.
    e. none of the above

RELATED ACTIVITIES

- Search the Internet to find information on cloning research. Compose a one-page essay summarizing what you have learned, and indicate specifically how cloning research may potentially be used to treat patients requiring organ transplants.
- List three issues associated with the use of DNA technology that pose potential ethical dilemmas or societal risks. What are these social and ethical concerns? Short of altogether prohibiting the use of new DNA manipulation techniques, what can be done to adequately control or regulate work in this area?
- Use the Internet or other resources to learn about the genetically modified plant dubbed "Golden Rice." Describe the arguments in favor of using genetically engineered rice to combat blindness in third-world countries. Do you believe this is a safe and effective approach to nutrition supplementation? State your reasons why or why not.

ANSWERS AND EXPLANATIONS

Factual Knowledge

1. b. As we saw in the chapter, new genes may be inserted into various food crops to enable them to produce essential nutrients that would otherwise be lacking. For more information, see Section 16.4, Genetic Engineering.
2. d. The fact that DNA is structured the same way in all known organisms means that similar methods can be used to study the hereditary material. For more information, see Section 16.1, *Revolutionary techniques are the foundation of DNA technology.*
3. a. Once isolated, DNA is cut into fragments using restriction enzymes, then the fragments separated using gel electrophoresis. To complete the process, the

biologist applies an electrical charge across the gel to begin separating the DNA fragments by size. For more information, see Section 16.7, *Enzymes are used to cut and join DNA*, and Figure 16.11, Restriction Enzymes and Gel Electrophoresis Can Be Used to Identify the Sickle-Cell Allele.

4. b. Recall that ligases are the enzymes used to glue together DNA fragments. For more information, see Section 16.7, *Enzymes are used to cut and join DNA*.

5. c. DNA probes are used in hybridization experiments. These short, single-stranded segments of DNA are used to identify the presence of target DNA sequences. For more information, see Section 16.7, *DNA sequencing and DNA synthesis are key tools in biotechnology*.

6. a. The determining factor in the movement of DNA through an electrophoresis gel is the size of the fragment. For more information, see Section 16.7, *Gel electrophoresis sorts DNA fragments by size*.

7. e. DNA hybridization involves the breaking of DNA into smaller fragments using restriction enzymes, heating the DNA to separate the strands, then probing for target sequences using labeled DNA probes. For more information, see Section 16.7, *DNA sequencing and DNA synthesis are key tools in biotechnology*.

8. d. The definition of a cloned gene is one that has been isolated and repeatedly duplicated. For more information, see Section 16.1, *Revolutionary techniques are the foundation of DNA technology*.

9. False. Cloning can be used to obtain more copies of an animal with desirable genetic traits, such as pigs that have been engineered to produce organs suitable for human transplants or cows that make pharmaceutical products in their milk. For more information, see Section 16.3, Reproductive Cloning of Animals.

10. d. Restriction enzymes are used to cut DNA at specific sites, which may be important in preparing the target DNA but is not required for the PCR process itself. For more information, see Section 16.7, *PCR is used to amplify small quantities of target DNA*.

11. b. Besides the gene gun, other methods of genetic transfer between cells or organisms include the use of vectors such as viruses and plasmids. For more information, see Section 16.4, Genetic Engineering.

12. e. DNA fingerprinting can be performed with virtually any source containing DNA, including the tissues and fluids mentioned in the question. For more information, see Section 16.2, DNA Fingerprinting.

13. e. Although DNA technology is often used to introduce foreign genes into an organism, it cannot be used to remove diseased genes from a multicellular organism because each cell in the body would need to be treated separately. For more information, see Section 16.4, Genetic Engineering, and Section 16.5, Human Gene Therapy.

14. a. Gene therapy, although showing great theoretical promise, is still in its infancy as a practical means of treating the cause of genetic disorders. The technique has met with limited success and a few disturbing failures, but it is important to note that the positive results have not been clearly attributable to gene therapy alone. For more information, see Section 16.5, Human Gene Therapy.

15. c. DNA probe
    e. clone
    a. plasmid
    d. recombinant DNA
    b. gene therapy
    For more information, see Section 16.1, *Revolutionary techniques are the foundation of DNA technology*; Section 16.5, Human Gene Therapy; and Section 16.7, *DNA sequencing and DNA synthesis are key tools in biotechnology* and *DNA cloning is a means of propagating recombinant DNA*.

Conceptual Understanding

1. e. All of the statements listed are concerns for critics of genetic engineering. Although some of these concerns may be alarming, since 1996 more than 75 percent of all processed foods in the United States contain ingredients derived from genetically engineered plants. For more information, see Section 16.6, Ethical and Social Dimensions of DNA Technology.

2. e. All of these choices are used to study DNA-based genetic systems on Earth, and they would likely be universal among other DNA systems as well. For more information, see Section 16.7, A Closer Look at Some Tools of DNA Technology.

3. False. Both ligases and polymerases are used to synthesize and assemble DNA sequences, not to break them down. For more information, see Section 16.7, *Enzymes are used to cut and join DNA* and *PCR is used to amplify small quantities of target DNA*.

4. a. Migration through the electrophoresis gel is a function of the size of the DNA fragments, with small fragments moving farthest, as they are able to "squeeze" through the gel more easily. For more information, see Section 16.7, *Gel electrophoresis sorts DNA fragments by size*.

5. c. Because it moved the shortest distance through the gel, DNA sequence *z* must be the largest. This is why we can conclude that *z* represents the sickle-cell allele, which cannot be cut by the restriction enzyme used in this test. The fragments *x* and *y* are smaller, produced when the normal allele is cut by the

enzyme. For more information, see Section 16.7, *Gel electrophoresis sorts DNA fragments by size.*

6. b. Because both probes bind, both of the target DNA sequences (alleles) must be present. The individual is therefore heterozygous. For more information, see Section 16.7, *DNA sequencing and DNA synthesis are key tools in biotechnology.*

7. e. All of these are possible once a gene has been cloned. For more information, see Section 16.7, *DNA cloning is a means of propagating recombinant DNA.*

8. a. For the cut organismal and plasmid DNA fragments to connect properly, they must have similar ends on their DNA sequences following digestion with restriction enzymes. This can be accomplished only by using the same restriction enzyme. Using different enzymes for each type of DNA would result in incompatible fragments with different sequences at their ends. For more information, see Section 16.7, *Enzymes are used to cut and join DNA.*

9. b. PCR is a way to dramatically increase the numbers of a particular sequence of DNA (or an entire gene). This technique is most useful when cloning DNA, because prior knowledge of the DNA sequence must first be obtained. For more information, see Section 16.7, *PCR is used to amplify small quantities of target DNA.*

10. b. Although DNA fingerprinting is a powerful technique, a genetic match can demonstrate only a high probability of guilt because there is a 1/100,000 to 1/1,000,000,000 chance that two individuals will share the same genetic profile. For more information, see Section 16.2, DNA Fingerprinting.

11. b. These are all good examples of the end products of various DNA technologies. For more information, see Section 16.1, The Brave New World of DNA Technology.

12. c. The use of BGH does not appear to have serious health or environmental drawbacks, but it could offer such a competitive edge to corporate farming operations that smaller, more traditional family farms might be forced out of business. This scenario is a good example of how genetic engineering can have unintended social consequences. For more information, see Section 16.6, Ethical and Social Dimensions of DNA Technology.

13. False. The application of gene therapy is still quite limited, and the few apparent successes are open to speculation about just how much the gene therapy actually accounted for the patient's improvement. For more information, see Section 16.5, Human Gene Therapy.

14. b. Genetically modified foods are found in many commercial products such as frozen pizzas, oils, and corn syrup. For more information, see the Biology Matters box, "Have You Had Your GMO Today?"

CHAPTER 17 | How Evolution Works

GETTING STARTED

Below are a few questions to consider before reading Chapter 17. These questions will help guide your exploration and assist you in identifying some of the key concepts presented in this chapter.

1. What was the name of the ship on which Charles Darwin sailed to the Galápagos Islands?

2. What household item did the American Medical Association tell U.S. consumers to avoid using in 2000?

3. What do a person's arm, a whale's fin, and a bat's wing all have in common?

4. What can the fossil record tell us about how Earth's environments have changed over time?

5. What are vestigial organs?

6. By what amount does the distance between South America and Africa increase each year?

7. How does understanding evolution help with agriculture, pharmaceutical design, and other areas of importance in everyday life?

A GUIDE TO THE READING

The following concepts typically give students the most difficulty when exploring the content in Chapter 17 for the first time. For each concept, one or more references have been identified that may help you gain a better understanding of these potentially problematic areas.

Mechanisms of Evolution

Evolution within a population occurs when individuals possessing certain characteristics survive and reproduce at a higher rate than other individuals. As described in the chapter, there are two primary means by which this occurs: natural selection and genetic drift. Although both mechanisms result in the evolution of a population, they accomplish this in very distinct ways. In natural selection, the characteristics an individual possesses actually contribute to the individual's survival and rate of reproduction. Because these characteristics may be inherited, subsequent generations would also possess these characteristics and also have higher survival and reproduction rates. Over several generations, this would cause the population to change to reflect the abundance of individuals with these favorable characteristics. In contrast, genetic drift occurs when natural events remove particular groups of individuals with particular characteristics by chance. As a result, surviving individuals with distinct characteristics go on to reproduce, passing their inheritable characteristics to the next generation. The key to understanding the difference between these mechanisms is realizing that in genetic drift, inheritable characteristics play no role in determining survival. Survival of individuals with particular characteristics occurs solely by chance.

For more information on this concept, be sure to focus on

- Section 17.2, Mechanisms of Evolution
- Figure 17.2, Darwin Proposed Natural Selection as the Mechanism behind Species Formation
- Figure 17.4, Genetic Drift Can Drastically Alter the Genetic Makeup of a Population

Shared Characteristics

As we saw in the chapter, different species may share many characteristics. Features that are shared between organisms that have a common ancestor are called "homologous." A good example would be the opposable thumbs shared by the great apes and humans. These two groups of organisms shared a recent common ancestor who likely also had this distinguishing feature. Analogous features, in contrast, appear when different groups of organisms develop similar characteristics separately through natural selection as a result of similar environmental pressures. This process is referred to as "convergent evolution." The example provided in the text indicates that the features shared by the distantly related organisms sharks and dolphins (such as large dorsal fins) would have evolved through convergent evolution. As a result, these features would be considered analogous.

For more information on this concept, be sure to focus on

- Section 17.3, Evolution Can Explain the Unity and Diversity of Life
- Figure 17.6, Shared Characteristics
- Figure 17.8, The Power of Natural Selection

TYING IT ALL TOGETHER

Several concepts presented in this chapter build on those presented in previous chapters and may also be revisited and discussed in greater detail in subsequent chapters, including

Evolutionary Trees and Branching

- Chapter 2—Section 2.1, *Evolutionary divergence explains the diversity of life on Earth*
- Chapter 2—Section 2.2, The Linean System of Biological Classification

Genes and Alleles

- Chapter 12—Section 12.2, Basic Patterns of Inheritance
- Chapter 18—Section 18.1, Alleles and Genotypes

Evolution of Populations

- Chapter 18—Section 18.2, Four Mechanisms That Cause Populations to Evolve

Mutation

- Chapter 12—Section 12.1, *Gene mutations are the source of new alleles*

- Chapter 13—Section 13.2, Origins of Genetic Differences between Individuals
- Chapter 18—Section 18.3, Mutation: The Source of Genetic Variation

Genetic Drift

- Chapter 18—Section 18.5, Genetic Drift: The Effects of Chance

Natural Selection

- Chapter 18—Section 18.6, Natural Selection: The Effects of Advantageous Alleles
- Chapter 18—Section 18.7, Sexual Selection: Where Sex and Natural Selection Meet

Speciation

- Chapter 19—Section 19.4, Speciation: Generating Biodiversity

Using the Fossil Record to Determine Evolutionary Relationships

- Chapter 20—Section 20.1, The Fossil Record: A Guide to the Past

PRACTICE QUESTIONS

Factual Knowledge

1. Darwin's *The Origin of Species* first appeared in
   a. 1831.
   b. 1836.
   c. 1859.
   d. 1900.
   e. 1953.

2. Which of the following scholars made the greatest positive contribution to Darwin's work?
   a. James Ussher
   b. George-Louis Leclerc de Buffon
   c. Jean-Baptiste Lamarck
   d. Reverend Thomas Malthus
   e. Aristotle

3. Biological evolution is best defined as
   a. the change in genetic characteristics of populations over time.
   b. random shifts in the genetic makeup of a population.
   c. changes in populations over time.

d. characteristics that improve the chances of survival for a particular organism.

e. allowing only individuals with certain characteristics to breed.

4. When natural selection is operating, which of the following is likely to occur?
   a. The characteristics of individuals will evolve over their lifetime.
   b. Individuals within a population may have greater survival than others.
   c. The evolution of a population is driven by random natural events.
   d. Mutations that occur in one cell are spread to the remainder of the cells in an organism.
   e. all of the above

5. Which of the following is an example of a shared characteristic between species that exists because of common evolutionary descent?
   a. the type of flower that bees will preferentially pollinate
   b. the number and type of bones in the limbs of vertebrates
   c. the similarity of color and pattern between frogs and their environment
   d. mealybugs' ability or inability to feed on genetically engineered corn
   e. all of the above

6. Modern evidence strongly supports the evolutionary prediction that organisms should possess signs of their common ancestry with other organisms. (True or False)

7. With which of the following statements would Darwin most likely disagree?
   a. Individuals within a population vary in the characteristics they possess.
   b. Evolution is best viewed as a purposeful and directed change over time.
   c. Natural selection is the mechanism by which biological evolution takes place.
   d. The fossil record supports the view that biological evolution has occurred.
   e. all of the above

8. Science has never actually observed the process of evolutionary change in species. (True or False)

9. Most religious leaders today view evolution and religion as compatible. (True or False)

10. The evidence supporting biological evolution includes
    a. fossils.
    b. patterns of embryological development.
    c. shared anatomical characteristics.
    d. molecular similarities.
    e. all of the above

11. Which of the following animals is a product of artificial selection?
    a. horse
    b. tuna
    c. dalmation
    d. medium ground finch
    e. corn mealybug

12. Refer to Figure 17.15 in your textbook. Which of the following do all of these birds have in common?
    a. They live in Africa.
    b. Darwin knew nothing about them.
    c. They are the product of artificial selection.
    d. All are descended from a common ancestor.
    e. Their origins can be explained by continental drift.

13. Which of the following does *not* have a direct effect on the process of biological evolution?
    a. techniques for determining the age of fossils
    b. differences in reproductive success
    c. variation in the inheritance of traits
    d. changes in the environment
    e. natural selection

14. Match each term with the best description.
    ___ adaptation
    ___ gene pool
    ___ natural selection
    ___ genetic drift
    ___ speciation
    a. all genetic information in a population
    b. random change in the genetics of a population
    c. splitting of one type of organism into others
    d. mechanism that drives biological evolution
    e. beneficial characteristic of an organism

Conceptual Understanding

1. Many bizarre plants and animals inhabit the Galápagos Islands, as young Charles Darwin observed when he visited there in the early 1830s. Which of the following is the most likely reason that oceanic islands often have such unusual inhabitants?
   a. Natural selection does not work on islands.
   b. Evolution takes place more slowly on islands than it does elsewhere.
   c. Islands are isolated and thus susceptible to strong natural selection.
   d. Islands usually have many more species than do mainland habitats.
   e. Artificial selection is more common on islands than on the mainland.

2. Among individuals, the ultimate source of heritable differences fueling natural selection is
   a. gene mutation.
   b. industrial pollution.
   c. continental drift.
   d. all of the above
   e. none of the above

3. The evolutionary effects of genetic drift are
   a. most obvious in large populations.
   b. usually random and unpredictable.
   c. always the same as those of natural selection.
   d. shaped by nonrandom selective forces.
   e. exemplified by population changes in the peppered moth.

4. With which of the following statements would an evolutionary biologist agree?
   a. The fossil record is the only source of evidence in favor of evolution.
   b. Evolving populations rarely show changes in their genetic makeup.
   c. There has been no significant change in organisms through time.
   d. Mutations provide the foundation for evolutionary change.
   e. God was the designer of all organisms in the form in which they now appear.

5. Anteaters and some whales lack teeth as adults. The presence of teeth in the embryos of these creatures is evidence that
   a. all animals must undergo some type of embryological development.
   b. the embryos have a very different diet from that of the adults.
   c. these species are the product of artificial selection.
   d. these species share a common ancestry with animals that have adult teeth.
   e. none of the above

6. The cytochrome c data in Figure 17.10 suggests that which of the following statements is true?
   a. Rhesus monkeys and humans are fairly closely related.
   b. Humans and moths are only distantly related.
   c. Tuna and dogs are only distantly related.
   d. all of the above
   e. none of the above

7. Again refer to Figure 17.10. If you were to plot the cytochrome c data for a frog on this graph, where would it most likely be placed in relationship to the other animals?
   a. between the human and the rhesus monkey
   b. between the rattlesnake and the dog

   c. between the tuna and the rattlesnake
   d. between the moth and the tuna
   e. to the left of the moth

8. Basing your answer on the information in Figure 17.6, which of the following organisms should have five digits and the same set of arm bones?
   a. chimpanzee
   b. cat
   c. squirrel
   d. dolphin
   e. all of the above

9. The evolution of horses, shown in Figure 17.9, is a good example of the evolutionary process because it
   a. provides solid fossil evidence in favor of evolution.
   b. lacks a number of key specimens and so is rather fragmented.
   c. shows that Earth is indeed only 10,000 years old.
   d. demonstrates a decrease in body size between *Hyracotherium* and *Equus*.
   e. none of the above

10. Consider two different species of turtle. These organisms
    a. are members of the same population.
    b. are capable of breeding with each other.
    c. share very few anatomical characteristics.
    d. were likely formed through natural selection.
    e. none of the above

11. The rudimentary hind legs of a python and the fully functional legs of other reptiles are
    a. shared characteristics that result from common descent.
    b. freaks of development that have nothing to do with evolution.
    c. the result of developmental abnormalities that sometimes occur.
    d. examples of fossil evidence in favor of evolution.
    e. not the products of natural selection.

12. The Galápagos Islands have played a significant role in the development of evolutionary thinking for more than 150 years. Which of the following is *not* an example of evolutionary change that has been studied using species from the Galápagos?
    a. short-term species adaptation to drought conditions
    b. breeding of new varieties from a single wild species
    c. results that occur when a new species finds an unused habitat
    d. effects of geographic isolation on the process of natural selection
    e. the effects of environmental pressures on the adaptations of individual species

RELATED ACTIVITIES

- Use your school library or the Internet to research Charles Darwin and his ideas. Compose a one-page essay about his educational background and how his scientific ideas were developed. Be sure to consider the historical context of society at the time Darwin introduced his ideas about evolution and how that societal perspective influenced the initial acceptance of *The Origin of Species*.
- Using the Internet, find an example other than horses that shows good fossil evidence for evolution. Compose a one-page summary explaining the example you found.
- Review the chapter's Biology Matters box, "Can't Live with 'Em, Can't Live without 'Em," which deals with the overuse of antibacterial products. Perform an Internet search for information on the evolution of antibiotic resistance and the problem this has caused for the medical community. Compose a one-page summary of your results, focusing on the role natural selection plays in this evolutionary phenomenon.

ANSWERS AND EXPLANATIONS

Factual Knowledge

1. c. Darwin set sail on the *Beagle* in 1831 and returned to England in 1836. He published *The Origin of Species* in 1859. For more information, see Section 17.1, *Darwin offers a unifying explanation: descent with modification.*
2. d. Reverend Malthus's observations on the growth of populations and the competition for resources led to the idea of differential reproduction, a key aspect of natural selection. For more information, see Section 17.1, *Darwin offers a unifying explanation: descent with modification.*
3. a. The idea of genetic change over time is key to understanding biological evolution. For more information, see Section 17.2, Mechanisms of Evolution.
4. b. One of the keys to natural selection is the enhanced survival of individuals with particular inheritable characteristics that can be passed on to subsequent generations. For more information, see Section 17.2, *Natural selection generates adaptations in a population.*
5. b. The simplest explanation is that this limb bone pattern is an ancestral characteristic that evolved at the beginning of the vertebrate lineage and thus formed the basis for the body shape of all groups that followed. For more information, see Section 17.3, *Organisms share characteristics as a result of common descent.*

6. True. Examples include internal anatomy, details of molecular structure, and embryological patterns. For more information, see Section 17.4, *Organisms contain evidence of their evolutionary history.*
7. b. There is nothing purposeful or directed about natural selection and its evolutionary outcomes. There is only a filtering of possible adaptations using the backdrop of environmental pressures that change on a regular basis. For more information, see Section 17.2, *Natural selection generates adaptations in a population.*
8. False. Scientists observe change in the genetic makeup of populations (that is, biological evolution) all the time. There are also examples of new plant and animal species that have appeared within the history of modern science. For more information, see Section 17.4, *Direct observation reveals genetic changes within species.*
9. True. For example, the Catholic Church has accepted the idea that evolution may explain the physical characteristics of humans but that religion is necessary to explain our spiritual characteristics. Some other Christian denominations take similar positions, as do leaders of many other faith communities. For more information, see Section 17.5, The Impact of Evolutionary Thought.
10. e. All of these areas contribute evidence in support of biological evolution. For more information, see Section 17.4, The Evidence for Biological Evolution.
11. c. Humans have carried out selective breeding on domestic dogs for thousands of years, resulting in the evolution of many varieties of the gray wolf subspecies known as "*Canis lupus familiaris*." For more information, see Section 17.4, *Direct observation reveals genetic changes within species*, and Figure 17.11, Artificial Selection Produces Genetic Change.
12. d. The Galápagos finches have evolved into a variety of forms, but all are descended from a single species that came from the mainland some time in the evolutionary past. For more information, see the chapter's Applying What We Learned article, "Darwin's Finches: Evolution in Action."
13. a. Although these techniques give us a better picture of the way fossil organisms are related to each other chronologically, they are not themselves a part of the evolutionary process. For more information, see Section 17.4, *Evolution is strongly supported by the fossil record.*
14. e. adaptation
    a. gene pool
    d. natural selection
    b. genetic drift
    c. speciation

For more information, see Section 17.1, *Darwin offers a unifying explanation: descent with modification*; Section 17.2, Mechanisms of Evolution, *Genetic drift generates differential reproduction through accidental events,* and *Natural selection generates adaptation in a population*; and Section 17.3, *The diversity of life results from the splitting of one species into two or more species.*

Conceptual Understanding

1. c. Because they are isolated, islands often have very specific selective pressures that can dramatically shape adaptations. For more information, see Section 17.2, Mechanisms of Evolution.

2. a. Gene mutation is the only way new alleles can form, and new alleles are what lead to new, and potentially adaptive, variations. For more information, see Section 17.2, *Mutations introduce genetic variation in a population.*

3. b. Genetic drift is the random and unpredictable change in the genetic makeup of a population as a result of small population sizes or chance events. For more information, see Section 17.2, *Genetic drift generates differential reproduction through accidental events.*

4. d. Heritable differences are required for evolution to occur. These heritable differences are generated through gene mutation. For more information, see Section 17.2, *Mutations introduce genetic variation in a population.*

5. d. The loss of teeth after embryological development suggests that these species are descended from toothed ancestors, and that they have subsequently lost their adult teeth as an adaptive response to selective pressures in their environment. For more information, see Section 17.4, *Organisms contain evidence of their evolutionary history.*

6. d. All of these are correct interpretations of the data based on the degree of similarity in cytochrome c

structure indicated. Species with the fewest mutations between them are closely related, whereas those with large differences are more distantly related. For more information, see Section 17.4, *DNA evidence provides some of the most compelling evidence for evolution.*

7. c. Frogs are amphibians and are intermediate between fish and reptiles in terms of evolutionary origins. Their cytochrome c structure would therefore reflect this evolutionary history. For more information, see Section 17.4, *Organisms contain evidence of their evolutionary history* and *DNA evidence provides some of the most compelling evidence for evolution.*

8. e. Because all are mammals, they too should show the same basic pattern of limb bones seen in humans, whales, and bats. For more information, see Section 17.3, *Organisms share characteristics as a result of common descent.*

9. a. Only some of the fossil evidence now available for horses is shown in Figure 17.9. Even without these additional data, it is clear how horses represent a steady progression from smaller to larger animals. For more information, see Section 17.4, *Evolution is strongly supported by the fossil record.*

10. d. To the best of our knowledge, natural selection is the mechanism by which virtually all new species arise. For more information, see Section 17.3, *The diversity of life results from the splitting of one species into two or more species.*

11. a. Because both of these species share some form of the feature, it must have been present in a common ancestor. For more information, see Section 17.3, *Organisms share characteristics as a result of common descent.*

12. b. This would be similar to the example of wild wolves that were domesticated and artificially bred to produce various dog breeds. Such studies have not been part of the Galápagos research history. For more information, see Section 17.4. *Direct observation reveals genetic changes within species.*

CHAPTER 18 | Evolution of Populations

GETTING STARTED

Below are a few questions to consider before reading Chapter 18. These questions will help guide your exploration and assist you in identifying some of the key concepts presented in this chapter.

1. How did the availability of antibiotics, such as penicillin, change medicine and health in America?

2. How has human use of pesticides like DDT influenced insect populations?

3. What causes the human immunodeficiency virus (HIV) to act like a "moving target" in an infected individual's body?

4. What do the Florida panther, northern elephant seal, and African cheetah all have in common?

5. What legislation has caused the directional selection of light-colored moths over dark-colored moths in England over the past several decades?

6. What is the ideal birth weight for human babies, as determined by stabilizing selection?

A GUIDE TO THE READING

The following concepts typically give students the most difficulty when exploring the content in Chapter 18 for the first time. For each concept, one or more references have been identified that may help you gain a better understanding of these potentially problematic areas.

Genetic Drift

Genetic drift is the process by which alleles are selected within a population, at random, over time. The frequency of particular alleles may change as a result of random events. This may have a profound effect on the population if the size of the population happens to be small. Because small populations have fewer alleles in total, if a number of individuals happen to leave or are eliminated from the population, the relative effect on the remaining alleles can be drastic. When one allele reaches a frequency of 100 percent as a result of the genetic drift of other alleles, the remaining allele is said to have reached "fixation"; that is, because this is the only allele present in the population, it will not change over time. It is important to note that the process of genetic drift is not subject to adaptive evolution; it may result in the fixation of alleles that are beneficial, neutral, or even harmful to the population.

For more information on this concept, be sure to focus on

- Section 18.5, *Genetic drift affects small populations*
- Figure 18.4, Genetic Drift Occurs by Chance

Types of Natural Selection

As we saw in the chapter, there are three different types of natural selection, each shaping the evolution of a population as a result of differences in the survival and reproductive success of members possessing particular traits: (1) Directional selection occurs when individuals possessing an extreme trait have a reproductive or survival advantage over individuals lacking this trait. An example of directional selection might include the neck length of giraffes. Giraffes with longer necks might have a survival advantage over animals with shorter necks because they may be more adept at feeding from tall trees. The characteristic determining neck length would then be inherited by subsequent generations, shifting the characteristic of the population to possession of longer necks. (2) Stabilizing selection occurs when individuals with an intermediate trait have a survival or reproductive advantage over individuals with extreme versions of the trait. The example provided in the chapter focuses on human birth weight; babies that are underweight, premature, or both have a survival disadvantage, whereas very large babies are likely to cause problems during delivery. These factors have stabilized the average birth weight of babies at around eight pounds. (3) Disruptive selection occurs when individuals possessing a trait at either extreme have a survival or reproductive advantage over individuals with intermediate forms of the trait. The example provided in the chapter focuses on beak size within a population of African seed cracker birds. The advantage of having either a large or a small beak is that it enhances feeding efficiency, allowing populations to feed on either hard or soft seeds. Another way to look at this is to realize that individuals possessing intermediate-sized beaks would be at a disadvantage because they would be inefficient at feeding on both types of seeds.

For more information on this concept, be sure to focus on

- Section 18.6, *There are three types of natural selection*
- Figure 18.7, Directional, Stabilizing, and Disruptive Selection Affect Phenotypic Traits Differently
- Figure 18.8, Directional Selection in the Peppered Moth
- Figure 18.9, Stabilizing Selection for Human Birth Weight
- Figure 18.10, Disruptive Selection for Beak Size

TYING IT ALL TOGETHER

Several concepts presented in this chapter build on those presented in previous chapters and may also be revisited and discussed in greater detail in subsequent chapters, including

Genetic Recombination

- Chapter 13—Section 13.3, Genetic Linkage and Crossing-Over

Mutations

- Chapter 14—Section 14.4, Repairing Replication Errors and Damaged DNA

Microevolution

- Chapter 17—Section 17.1, Descent with Modification

Genetic Drift and Natural Selection

- Chapter 17—Section 17.2, Mechanisms of Evolution

Adaptation

- Chapter 19—Section 19.1, Adaptation: Adjusting to Environmental Challenges

Speciation

- Chapter 19—Section 19.4, Speciation: Generating Biodiversity

PRACTICE QUESTIONS

Factual Knowledge

1. Which of the following is an example of rapid evolutionary change in a population?
   a. genetic bottlenecks in prairie chickens
   b. sexual selection in gophers and pine trees
   c. pesticide resistance in insects
   d. average birth weight in humans
   e. all of the above

2. "Microevolution" refers to
   a. the evolution of microorganisms.
   b. evolution that occurs on a large scale.
   c. the changes in allele or genotype frequencies that occur in a population over time.
   d. the splitting of a large group of organisms into multiple species.
   e. none of the above

3. If there are 1,000 individuals in a population, and 600 of them are homozygous dominant for a gene that has only two alleles, what is the genotype frequency for the homozygous dominant condition?
   a. 1.0
   b. 0.6
   c. 0.5
   d. 0.4
   e. 0.3

4. A particular gene in a population has two alleles, C and c. If the allele frequency of $C = 0.7$, what is the frequency of c?
   a. 1.0
   b. 0.7
   c. 0.3
   d. 0.0
   e. cannot determine from the information given

5. Which of the following statements about genetic variation is true?
   a. Mutation is the ultimate source of new alleles.
   b. The only source of genetic variation is mutation.
   c. New alleles are formed during sexual reproduction.
   d. Mutations that result in the production of new alleles are quite common.
   e. all of the above

6. The Hardy–Weinberg equation describes the genotype frequencies that will occur in an evolving population of organisms. (True or False)

7. In the Hardy–Weinberg equation, the terms "p" and "q" represent the _____ in a population.
   a. size of the gene pool
   b. total number of alleles
   c. genotype frequencies
   d. allele frequencies
   e. mutation rates

8. A violation of which of the following conditions would cause a population to shift away from its stable Hardy–Weinberg frequencies?
   a. no genetic drift
   b. no gene flow
   c. no net mutation
   d. no natural selection
   e. any of the above

9. Sexual selection in a population
   a. is a special type of natural selection.
   b. always favors individuals with a higher survival potential.
   c. alters allele frequencies in a stable population.
   d. favors individuals with poor reproductive potential.
   e. reduces genetic variation in the population.

10. Which of the following statements about gene flow is true?
    a. It reduces the genetic variation within populations.
    b. It reduces the genetic variation between populations.
    c. It reduces the movement of alleles between populations.
    d. It does not violate the conditions of the Hardy–Weinberg model.
    e. It rarely occurs in natural populations.

11. Which of the following is more likely to occur when a population of organisms is small?
    a. natural selection
    b. mutation
    c. heterozygote advantage
    d. genetic drift
    e. random mating

12. Of the organisms listed below, which one is *not* believed to have undergone a genetic bottleneck at some point in its recent evolutionary history?
    a. Culex mosquitoes
    b. Northern elephant seals
    c. Illinois prairie chickens
    d. African cheetahs
    e. Florida panthers

13. Changes in the coloration of pepper moths over time are one dramatic example of the mode of natural selection known as
    a. disruptive selection.
    b. directional selection.
    c. stabilizing selection.
    d. heterozygote advantage.
    e. genetic drift.

14. Match each term with the best description.
    __ allele frequency
    __ gene flow
    __ genotype frequency
    __ microevolution
    __ stabilizing selection
    a. p^2 in the Hardy–Weinberg equation
    b. short-term changes in allele frequencies within a population
    c. shifts toward the mean in a population
    d. proportion of a particular allele in the gene pool
    e. occurs as a result of migration between populations

Conceptual Understanding

1. The genetic mutations that occur in organisms are
   a. relatively rare events that have little consequence for evolution.
   b. predictable by scientists before they actually occur.
   c. directed toward a particular adaptive goal.
   d. the raw material of evolution.
   e. none of the above

2. The impact a mutation has on a population depends largely on the environment in which the mutation occurs. (True or False)

3. Consider a population that has reached its Hardy–Weinberg equilibrium frequencies. If this population is suddenly subjected to strong directional selection, both the allele and the genotype frequencies are likely to undergo change as a result. (True or False)

4. Consider a population with a total of 500 alleles for a specific gene. How many individuals are in this population?
   a. 1,000
   b. 500

c. 250
d. 125
e. cannot determine from the information given

5. A population of organisms has a gene for which there are two alleles, *D* and *d*. The allele frequency of *D* = 0.8. If this population satisfies all five of the Hardy–Weinberg conditions, what are the genotype frequencies that are expected in the next generation?
    a. *DD* = 0.04; *Dd* = 0.32; *dd* = 0.64
    b. *DD* = 0.64; *Dd* = 0.32; *dd* = 0.64
    c. *DD* = 0.04; *Dd* = 0.64; *dd* = 0.32
    d. *DD* = 0.64; *Dd* = 0.32; *dd* = 0.04
    e. cannot determine from the information given

6. Several generations into the future, a population of birds currently undergoing stabilizing selection will look
    a. just like it does now.
    b. more like the current mean.
    c. more like one of the extremes.
    d. more like both of the extremes.
    e. none of the above

7. In the Hardy–Weinberg equation, the term "2pq" represents the
    a. overall gene frequency of the population.
    b. frequency of both homozygous genotypes.
    c. frequency of the heterozygous genotype.
    d. allele frequencies of the population.
    e. effect of disruptive selection on the population.

8. Refer to Figure 18.4 in your textbook. If the population of wildflowers in this hypothetical simulation of genetic drift were made up of 1,000 individuals instead of 10, how would this affect the change in allele frequencies?
    a. There would probably be a much smaller change in allele frequencies.
    b. There would definitely be a much greater change in allele frequencies.
    c. There would be no difference between the two types of simulation.
    d. The *aa* genotype would come to dominate by generation III.
    e. The *Aa* genotype would come to dominate by generation III.

9. A population susceptible to genetic bottlenecks would have which of the following characteristics?
    a. small population size
    b. high mutation rates
    c. high levels of gene flow
    d. sexual selection
    e. evolution by natural selection

10. Which of the following would *not* be considered one of the steps in the evolutionary process?
    a. Random mutations and genetic rearrangements occur.
    b. Natural selection acts on the genetic variation present in a population.
    c. Inheritable characteristics are produced by random genetic events such as mutation.
    d. Allele frequencies in a population change over time.
    e. none of the above

11. In Figure 18.9, the survival of newborn humans is compared against their birth weight. Note that these data are from London and span the period between 1935 and 1946. If you were to look at a graph of comparable data for the same population from the 1990s, what differences, if any, would you see?
    a. There would be no differences.
    b. There would be a broader optimum birth weight as a result of medical advances.
    c. There would be a sharper optimum peak as a result of medical advances.
    d. The range of weights over which newborns survive would have decreased.
    e. The length of time that underweight newborns survive would have decreased.

12. Which of the following contributes to the spread of antibiotic resistance among pathogenic bacteria?
    a. the transfer of resistance genes between bacteria
    b. poor sanitation in areas where pathogenic microbes are plentiful
    c. indiscriminate use of antibiotics
    d. directional selection
    e. all of the above

RELATED ACTIVITIES

• Search your library or the Internet to learn more about the problem of antibiotic resistance, which is described in the chapter. Compose a one-page summary of the results you have identified and propose a course of action our society should take to help control this growing problem. Be sure to focus on the economic and political factors that must be taken into account when addressing this problem.

• In the battle over evolution vs. creationism, one of the biggest problems is the comparison of microevolution with macroevolution. Your textbook has provided abundant evidence that microevolution occurs, as most creationists would acknowledge. However, creationists object to the idea that macroevolution naturally follows from microevolutionary change. What is your view about this debate? Is it reasonable to think that significant changes in species type can emerge from the accumulated small changes in allele frequencies that occur between populations? You may want to search the Internet at both pro-creation and pro-evolution sites to gather more information when developing your opinion.

- Consider the example provided in the chapter for directional selection in the Peppered Moth. Using the Internet, identify at least one other example of directional selection that has been influenced by the activities of humankind. Using your example, compose a one-page essay indicating how humankind has influenced the evolution of the species cited and what steps could be taken to restore the population to its original composition.

ANSWERS AND EXPLANATIONS

Factual Knowledge

1. c. The gene for pesticide resistance in mosquitoes is thought to have spread throughout major populations of these insects in only a few decades. All the other answers, while related to evolution, are not examples of rapid evolutionary change. For more information, see Section 18.3, Mutation: The Source of Genetic Variation.

2. c. Microevolution refers to the changes in allele or genotype frequencies that represent the smallest scale at which evolution can occur. For more information, see the chapter Introduction, Is Evolution Too Slow to Be Observed?

3. b. The genotype frequency is simply the proportion of that genotype in the population, in this case 600/1,000=0.6. Because the genotype frequencies must add up to 1.0 (all genotypes), the remaining 0.4 is made up of the heterozygotes and homozygous recessive individuals. For more information, see Section 18.1, Alleles and Genotypes.

4. c. Just like genotype frequencies, allele frequencies must add up to equal 1.0 (all alleles). Therefore, 1.0 − 0.7 = 0.3. For more information, see Section 18.1, Alleles and Genotypes.

5. a. Although mutation is the only source of new alleles, it is not the only source of genetic variation. Sexual recombination also produces variation, but it does not cause mutation. Finally, mutation rates generally are low in most populations. For more information, see Section 18.3, Mutation: The Source of Genetic Variation.

6. False. The Hardy–Weinberg equation describes the situation of unchanging genotype and allele frequencies and is valid only when a population is not evolving. For more information, see the chapter's Biology

Matters box, "Testing Whether Evolution Is Occurring in Natural Populations."

7. d. These allele frequencies can be used in the Hardy–Weinberg equation to generate expected genotype frequencies for the next generation of a nonevolving population. For more information, see the Biology Matters box, "Testing Whether Evolution Is Occurring in Natural Populations."

8. e. All of these conditions are necessary for the validity of the predictions generated by the Hardy–Weinberg equation. For more information, see the Biology Matters box, "Testing Whether Evolution Is Occurring in Natural Populations."

9. a. Recall that sexual selection occurs when a characteristic or set of characteristics makes individuals better at finding mates. As a result, these characteristics may be more readily inherited by subsequent generations. However, some of the characteristics that allow for better mating potential may actually be detrimental. For more information, see Section 18.7, Sexual Selection: Where Sex and Natural Selection Meet.

10. b. Gene flow moves alleles between populations, thus making those populations more alike genetically. For more information, see Section 18.4, Gene Flow: Exchanging Alleles between Populations.

11. d. Genetic drift is the random change in allele frequencies that occurs when a population is small. It stems from the strong influence of chance events when sample sizes are small. For more information, see Section 18.5, *Genetic drift affects small populations.*

12. a. Although the data for panthers, cheetahs, and elephant seals are not as clear-cut as the data for prairie chickens, these animals probably did suffer relatively recent genetic bottlenecks. Mosquitoes are interesting for other population-level evolutionary events. For more information, see Section 18.5, *Genetic bottlenecks can threaten the survival of populations.*

13. b. Poor air quality in Manchester, England, provided a selective advantage to darker-colored moths, which were better camouflaged. As air standards improved, lighter-colored moths had an advantage and were selected for. For more information, see Section 18.6, *There are three types of natural selection.*

14. d. allele frequency
   e. gene flow
   a. genotype frequency
   b. microevolution
   c. stabilizing selection
   For more information, see the chapter Introduction, Is Evolution Too Slow to Be Observed; Section 18.1, Alleles and Genotypes; the Biology Matters box, "Testing Whether Evolution Is Occurring in Natural

Populations"; Section 18.4, Gene Flow: Exchanging Alleles between Populations; and Section 18.6, *There are three types of natural selection.*

Conceptual Understanding

1. d. Although relatively rare, mutations are critical to evolution because they are the ultimate source of new alleles. They are not directed, nor are they predictable. For more information, see Section 18.3, Mutation: The Source of Genetic Variation.

2. True. The benefit or disadvantage of a particular mutation depends largely on the way the environment in which an organism lives "interprets" the phenotypic effect of the genetic change. For more information, see Section 18.3, Mutation: The Source of Genetic Variation.

3. True. Recall that populations in Hardy–Weinberg equilibrium will remain in equilibrium in the absence of mutation, gene flow, genetic drift, or selection. A population experiencing strong directional selection violates these "rules," and therefore a change in allele and genotype frequencies will result until equilibrium is reestablished. For more information, see the Biology Matters box, "Testing Whether Evolution Is Occurring in Natural Populations."

4. c. Because each diploid individual has two alleles, the number of individuals is half the total number of alleles in the population. For more information, see Section 18.1, Alleles and Genotypes.

5. d. We are told that the allele frequency for $D = 0.8$; therefore, the frequency of $d = 0.2$. These values are the "p" and "q" that we need to calculate the genotype frequencies in the next generation. Using the Hardy–Weinberg equation, p^2 (*DD*) $= 0.64$, $2pq$ (*Dd*) $= 0.32$, and q^2 (*dd*) $= 0.04$. For more information, see the Biology Matters box, "Testing Whether Evolution Is Occurring in Natural Populations."

6. b. Stabilizing selection shifts the population extremes toward the mean. For more information, see Section 18.6, *There are three types of natural selection.*

7. c. The terms "p" and "q" are the allele frequencies for a given gene, so "2pq" represents the two possible ways that heterozygotes can form. For more information, see the Biology Matters box, "Testing Whether Evolution Is Occurring in Natural Populations."

8. a. Genetic drift is a product of chance events that affect small populations much more than large populations. Therefore, in a population of 1,000 wildflowers, we would likely not see much change in allele frequencies due to genetic drift. For more information, see Section 18.5, *Genetic drift affects small populations.*

9. a. As with genetic drift, bottlenecks are a product of small population sizes. For more information, see Section 18.5, *Genetic bottlenecks can threaten the survival of populations.*

10. e. All of the events listed can be considered steps in the evolutionary process. For more information, see Section 18.2, Four Mechanisms That Cause Populations to Evolve; Section 18.3, Mutation: The Source of Genetic Variation; and Section18.6, Natural Selection: The Effects of Advantageous Alleles.

11. b. Medical advances now help underweight babies survive and allow large babies to be delivered surgically. Thus the stabilizing selection of natural childbirth has been lessened, which makes the curve less peaked. For more information, see Section 18.6, *There are three types of natural selection.*

12. e. The emergence of antibiotic-resistant strains of bacteria is due to the strong directional selection forces that act when antibiotics are used indiscriminately. The genes that confer resistance have been shown to spread among populations of bacteria. For more information, see the chapter's Applying What We Learned article, "Flesh-Eating Bacteria and Antibiotic Resistance."

Speciation and the Origins of Biological Diversity

GETTING STARTED

Below are a few questions to consider before reading Chapter 19. These questions will help guide your exploration and assist you in identifying some of the key concepts presented in this chapter.

1. How have so many species of cichlid fish in Lake Victoria evolved from only two ancestral species?

2. What unique adaptations of the "four-eyed fish" have enabled this species to survive in its environments?

3. What is the biological species concept?

4. In terms of speciation, how does reproductive isolation differ from geographic isolation?

5. What is a ring species?

6. Which animal species have been known to develop as a result of polyploidy?

7. What can the fossils of "hobbit" people in Indonesia teach us about geographic isolation?

A GUIDE TO THE READING

The following concepts typically give students the most difficulty when exploring the content in Chapter 19 for the first time. For each concept, one or more references have been identified that may help you gain a better understanding of these potentially problematic areas.

Reproductive Isolation

As defined in the chapter, the biological species concept indicates that a species is a group of organisms that are capable of interbreeding with each other. Different species remain as separate groups because they are not able to reproduce with members of other species. This can be because of either geographic isolation or reproductive isolation. Individuals from different species are usually incapable of interbreeding; therefore, they do not exchange genetic information, and they remain genotypically and phenotypically distinct. Species are said to be reproductively isolated from other species as a result of these reproductive barriers. Reproductive barriers can take several forms, including those that prevent successful mating (because of ecological, behavioral, mechanical, or gametic isolation), and those that prevent the survival or the success of hybrid species (zygote death, hybrid performance) when mating between two species occurs.

For more information on this concept, be sure to focus on

- Section 19.3, *Species are reproductively isolated from one another*
- Table 19.1, Barriers that Can Reproductively Isolate Two Species in the Same Geographic Region

Allopatric vs. Sympatric Speciation

Speciation occurs when a single species evolves into two or more distinct species over a period. The two main mechanisms by which speciation occurs are allopatric speciation and sympatric speciation. The key to understanding the difference between these mechanisms is to realize that allopatric speciation occurs when new species are formed as the result of the geographic separation and isolation of populations of the original species. Over time, these separated populations may evolve independently and become distinct in both genotype and phenotype. It is important to note that the distance separating the two populations must be great enough to inhibit gene flow between the populations. This factor makes the distance needed to achieve geographic separation vary, depending on the species (for example, organisms capable of flight will be able to cross new rivers or canyons easily). Sympatric speciation is the most common mechanism by which species form. It is important to note that sympatric speciation does not involve the geographic isolation of groups. Instead, other factors—such as polyploidy (multiple sets of chromosomes) or the natural selection of alleles that favor the use of one food source over another—may result in the formation of distinct species within the same geographic region.

For more information on this concept, be sure to focus on

- Section 19.4, *Speciation can result from geographic isolation*
- Section 19.4, *Speciation can occur without geographic isolation*
- Figure 19.11, Physical Barriers Can Produce Speciation by Blocking Gene Flow

TYING IT ALL TOGETHER

Several concepts presented in this chapter build on those presented in previous chapters and may also be revisited and discussed in greater detail in subsequent chapters, including

Meiosis and Chromosome Number (Ploidy)

- Chapter 10—Section 10.5, Meiosis: Halving the Chromosome Set to Make Gametes

Genetic Control of Development

- Chapter 33—*Section 33.6, Cell fate is controlled by differential gene expression*

Evolution

- Chapter 17—How Evolution Works

Gene Flow

- Chapter 18—Section 18.4, Gene Flow: Exchanging Alleles between Populations

Natural Selection

- Chapter 18—Section 18.6, Natural Selection: The Effects of Advantageous Alleles

PRACTICE QUESTIONS

Factual Knowledge

1. Your textbook's discussion of the many species of cichlids living in Lake Victoria illustrates that
   a. natural selection can improve the match between organism and environment.
   b. one sex can drive reproductive isolation of species.
   c. the environment can influence the success of a particular species.
   d. natural selection can operate quite rapidly.
   e. all of the above

2. The selective pressures that have served to increase the number of cichlid species in Lake Victoria are
   a. predation and feeding.
   b. predation and mate attraction.
   c. mate attraction and feeding.
   d. feeding and making nests.
   e. all of the above

3. Which of the following statements about adaptations is *false*?
   a. Adaptations provide a close match between the organism and its environment.
   b. Many adaptations appear complex, even though they are caused by simple mechanisms.
   c. All adaptations help the organism accomplish important functions.
   d. Examples of adaptations are difficult to find in the natural world.
   e. Some adaptations are associated with improved reproduction.

4. The degree of genetic variation present in a species is one of the factors that potentially limits the evolutionary effect of an adaptation. (True or False)

5. Developmental limitations on adaptation are due in part to
   a. geographic and reproductive isolation.
   b. the unintended effects of adaptive evolution.
   c. the dramatic effects on phenotype caused by altering developmental genes.
   d. whether the individual reproduced during the previous year.
   e. all of the above

6. Because adaptations are often compromises between conflicting ecological pressures, most adaptations will likely be
   a. poor matches with the environment.
   b. good matches with the environment, although not perfect.
   c. excellent matches with the environment, and nearly perfect.

d. absolutely perfect matches with the environment.

e. unaffected by adaptive evolution.

7. Figure 19.7 in your textbook provides an example of
   a. conflicting selective pressures that limit adaptation.
   b. the similarities in adaptation within a population of organisms.
   c. geographic isolation and the formation of hybrids.
   d. reproductive isolation due to a lack of gene flow.
   e. all of the above

8. According to your textbook's definition of the biological species concept, the two types of oak pictured in Figure 19.8 are members of the same species. (True or False)

9. Which of the following is *not* a barrier that can lead to reproductive isolation between species living in the same geographic region?
   a. behavioral isolation
   b. zygote death
   c. hybrid performance
   d. hybridization
   e. ecological isolation

10. A definition of the term "species" that is based on reproduction has distinct limitations, which include its being
    a. of little use when defining fossil species.
    b. inapplicable to species that use asexual reproduction.
    c. of little use when organisms produce hybrids.
    d. unclear and imprecise.
    e. all of the above

11. Which of the following cannot occur when a population is in reproductive isolation from others of its kind?
    a. gene flow
    b. mutation
    c. natural selection
    d. nonrandom mating
    e. genetic drift

12. Which of the following groups demonstrates the formation of ring species as a result of geographic isolation?
    a. birds of New Guinea
    b. salamanders of California
    c. plants in Hawaii
    d. cichlids in Africa
    e. all of the above

13. Speciation in animals can occur in the absence of geographic isolation. (True or False)

14. Match each term with the best description.
    ___ allopatric speciation
    ___ coevolution
    ___ polyploidy
    ___ speciation
    ___ hybrid
    ___ sympatric speciation
    a. multiple sets of chromosomes
    b. results from a cross between related species
    c. a major cause of living diversity
    d. occurs without geographic isolation
    e. adaptations that occur in tandem
    f. reproductive isolation is needed to get this

Conceptual Understanding

1. In terms of species, what do the weaver ant and moth caterpillar examples discussed in your textbook have in common?
   a. The behavior and morphology of both are the result of natural selection.
   b. They are both the result of recent speciation events.
   c. Neither species existed before the 1950s.
   d. Both are examples of domesticated species.
   e. They are both adapted for life on coastal sand dunes.

2. If a murky lake is stocked with a number of species of cichlid fish from Lake Victoria, what is most likely to happen?
   a. The cichlid species will be very visible to the many predators in this area.
   b. The male cichlid will be very attractive to females in the lake.
   c. The cichlid species will interbreed because the lake is so cloudy.
   d. The cichlid species will have many offspring because they are so colorful.
   e. none of the above

3. If a guppy species found in a stream is placed in a lake, its color will evolve rapidly from colorful to drab over as few as 10 generations because of the
   a. lack of food.
   b. presence of additional predators in the lake.
   c. temperature of the air.
   d. taste of the water.
   e. none of the above

4. Consider the caterpillars pictured in Figure 19.2. These two organisms are
   a. members of the same species.
   b. matched well with their environment.
   c. influenced developmentally by chemicals in their food.

d. the larval stage of a moth.

e. all of the above

5. Reproductive isolation can occur in a number of ways, but the net effect is always the same in that
   a. hybridization must take place before new species can form.
   b. it causes geographic isolation between organisms.
   c. the rate of speciation slows down.
   d. few or no genes flow between populations.
   e. females survive longer than males.

6. When different species mate and produce fertile offspring, those offspring
   a. are automatically considered new species.
   b. are said to be the product of hybridization.
   c. usually are more successful than their parents.
   d. show the characteristics of only one parent.
   e. all of the above

7. Which of the following must take place for speciation to occur?
   a. hybridization
   b. geographic isolation
   c. polyploidy
   d. reproductive isolation
   e. all of the above

8. Geographic isolation is said to occur whenever populations are separated by a distance that is great enough to limit
   a. hybridization.
   b. gene flow.
   c. polyploidy.
   d. domestication.
   e. none of the above

9. Reproductive isolation is to speciation as natural selection is to
   a. adaptive evolution.
   b. hybridization.
   c. ecological isolation.
   d. polyploidy.
   e. building an ant nest.

10. Polyploidy is believed to be in some way responsible for the origin of more than half of the animal species alive today. (True or False)

11. Although there have been many possible causes identified for the disappearance of species, foremost among the current threats to biodiversity is the
    a. possibility of a collision with a large asteroid.
    b. effect of climate change.
    c. destruction of habitats by humans.
    d. damage done by volcanic eruptions.
    e. none of the above

12. If you were to randomly select two species that have a common ancestor and measure the time it took for them to diverge from the ancestor, your result would likely be
    a. months.
    b. a few years.
    c. hundreds of years.
    d. a few thousand years.
    e. millions of years.

13. Although populations of North American apple maggot flies are not geographically isolated, they appear to be in the process of diverging into two new species because the two populations
    a. feed on two different types of food.
    b. reproduce at different times.
    c. lay their eggs on specific food types.
    d. have very little gene flow between them.
    e. all of the above

14. The distance needed for geographic isolation is the same for a crow as it is for a snail. (True or False)

15. Which of the following is *not* a case where adaptation is limited by other factors?
    a. An insecticide is highly effective at killing mosquitoes because there is no gene for resistance.
    b. Soapberry bugs have shorter beaks after introduction of a new food source.
    c. Peacocks have poor camouflage because peahens prefer bright-colored feathers.
    d. Larval insects lack compound eyes that allow adult insects to avoid predators.
    e. Hormones that increase female fertility also increase risk of disease.

RELATED ACTIVITIES

• Refer to the portion of Section 19.4 about the extinct "hobbit people" discovered in Indonesia. Consider how geographic or reproductive isolation (or both) likely contributed to the formation of this human species. Compose a one-page analysis of how you believe *Homo sapiens* and *Homo floresiensis* might have coexisted today, providing the latter species was still present. Assuming the two species were able to produce hybrids, what other factors might have contributed to the reproductive isolation of the two groups?

• Domestication is a special case of artificial selection in which humans select for specific adaptations that make an animal desirable for human purposes. Using your library or the Internet, identify an example of animal domestication. Compose a one-page summary discussing how and when the domestication of this animal took place and what changes in the species have resulted from human-controlled breeding programs. Pay particular attention to any negative traits that may have been introduced as trade-offs in the selection of desirable traits.

- Search the Internet for an example of a species that scientists believe may currently be in the process of speciation. In a one-page essay, describe the biology of this species, and discuss why scientists think that it may be about to give rise to multiple species.

ANSWERS AND EXPLANATIONS

Factual Knowledge

1. e. The effect of male cichlid body coloration and female mate choice on reproduction, along with the difference in jaw structure among the immense number of species and the presence of predators and pollution in Lake Victoria, illustrates all of these points. For more information, see the chapter's Applying What We Learned article, "Lake Victoria: Center of Speciation."

2. c. The tremendous number of species evolved because of mate selection and the evolution to specialize on various food items. Recall that predation by the Nile perch reduces the number of species. For more information, see the Applying What We Learned article, "Lake Victoria: Center of Speciation."

3. d. Adaptations that match the organism to its environment are everywhere in nature. For more information, see Section 19.1, Adaptation: Adjusting to Environmental Challenges.

4. True. If adaptive evolution is going to influence a trait, there must be some degree of variation in the trait upon which natural selection can work. For more information, see Section 19.2, *Lack of genetic variation can limit adaptation.*

5. c. Because relatively few genes control much of development, changes in those genes can have dramatic, and often negative, effects on development. This limits the degree to which organisms can change through adaptive evolution. For more information, see Section 19.2, *The varied effects of developmental genes can limit adaptation.*

6. b. Because of these conflicting pressures, most adaptations are ecological compromises that match organisms with their environment well but not perfectly. For more information, see Section 19.2, *Ecological trade-offs can limit adaptation.*

7. a. In the case of the male túngara frogs, the conflicting pressures are reproduction and predation. For more information, see Section 19.2, *Ecological trade-offs can limit adaptation.*

8. True. This example reveals one of the problems with defining species on the basis of reproduction, because such a definition does not easily accommodate species such as these oaks, which can form fertile hybrids. For more information, see Section 19.3, *Species are reproductively isolated from one another.*

9. d. Although poor reproductive performance by hybrids can be an isolating mechanism, the mere production of hybrids actually links the species reproductively. For more information, see Section 19.3, *Species are reproductively isolated from one another.*

10. e. All of these are potential problems with defining species on the basis of reproduction. For more information, see Section 19.3. *Species are reproductively isolated from one another.*

11. a. By definition, "reproductive isolation" means that there can be no flow of genes between the populations involved. For more information, see Section 19.3, *Species are reproductively isolated from one another.*

12. b. Recall that ring species occur when populations loop around a geographic barrier, such as a mountain. Populations at either of the extreme ends of the barrier, although capable of interbreeding with members of intermediate populations, are incapable of interbreeding with each other. For more information, see Section 19.4, *Speciation can result from geographic isolation.*

13. True. The East African cichlid fish discussed in this chapter are examples. For more information, see Section 19.4, *Speciation can occur without geographic isolation.*

14. f. allopatric speciation
    e. coevolution
    a. polyploidy
    c. speciation
    b. hybrid
    d. sympatric speciation
    For more information, see Section 19.1, *Adaptations can take many forms*; Section 19.3, *Species are reproductively isolated from one another*; and Section 19.4, Speciation: Generating Biodiversity, *Speciation can result from geographic isolation*, and *Speciation can occur without geographic isolation.*

Conceptual Understanding

1. a. Both are examples of the way natural selection shapes adaptive evolution, often in complex ways. For more information, see Section 19.1, *Adaptations can take many different forms.*

2. c. Because female cichlids rely on coloration to choose their mates, they will begin to interbreed. The cloudy water would prohibit them from distinguishing their own species on the basis of coloration. For more

information, see the Applying What We Learned article, "Lake Victoria: Center of Speciation."

3. b. The guppies found in the stream can be colorful because there are few predators. In a lake, there are more predators, and colorful individuals will stand out. Therefore, the color of the transplanted fish will change from bright to drab relatively rapidly. For more information, see Section 19.1, *Populations can adjust rapidly to environmental change.*

4. e. All of these are true statements about the remarkable adaptation exhibited by these caterpillars. For more information, see Section 19.1, *Adaptations can take many different forms.*

5. d. "Reproductive isolation" means that gene flow does not occur, and this sets the stage for possible divergence of traits that eventually can lead to speciation. For more information, see Section 19.3, *Species are reproductively isolated from one another.*

6. b. The ability of some species to reproduce fertile offspring through hybridization points out one of the limitations of a definition of species that is based on reproduction. For more information, see Section 19.3, *Species are reproductively isolated from one another.*

7. d. Some of the other answers listed are frequently associated with speciation, but only reproductive isolation is essential. For more information, see Section 19.4, Speciation: Generating Biodiversity.

8. b. The exact distance that constitutes geographic isolation of populations depends on the type of locomotion used by a species. This distance must limit the gene flow between isolated populations. For more information, see Section 19.4, *Speciation can result from geographic isolation.*

9. a. Reproductive isolation is necessary for speciation to occur. Likewise, natural selection must take place for adaptive evolution to occur. For more information, see Section 19.4, Speciation: Generating Biodiversity.

10. False. Polyploidy occurs with considerable frequency in plants: more than half of all living plant species are descended from ancestors that underwent polyploid speciation. This mechanism is used infrequently in animals. For more information, see Section 19.4, *Speciation can occur without geographic isolation.*

11. c. Although each of the other factors listed can adversely influence biodiversity, nothing has a greater impact on species numbers than the deterioration or destruction of habitats resulting from human activity. For more information, see the chapter's Biology Matters box, "How We Affect the Evolution of Others."

12. e. Although some natural speciation events can occur relatively rapidly (for example, over hundreds of or a few thousand years), most appear to take hundreds of thousands or even millions of years to complete. For more information, see Section 19.5, Rates of Speciation.

13. e. All of the above have contributed to the sympatric speciation of these flies. Although they are not geographically isolated, they have become reproductively isolated over time. For more information, see Section 19.4, *Speciation can occur without geographic isolation.*

14. False. The distance needed for populations to become geographically isolated depends on the mode of locomotion of the organisms. Animals such as birds can travel much greater distances over a day than can snails and therefore are more capable of reaching distant populations, facilitating gene flow. For more information, see Section 19.4, *Speciation can result from geographic isolation.*

15. b. Soapberry bugs were able to adapt rapidly to take advantage of a new food source, whereas all of the other examples are cases in which an animal population is unable to fully adapt to its conditions as a result of other factors, such as competing selective pressures or genetic limitations. For more information, see Section 19.2, Adaptation Does Not Craft Perfect Organisms.

CHAPTER 20

The Evolutionary History of Life

GETTING STARTED

Below are a few questions to consider before reading Chapter 20. These questions will help guide your exploration and assist you in identifying some of the key concepts presented in this chapter.

1. How many species of flowering plants are found on the continent of Antarctica?

2. How do fossils form?

3. Why are scientists so interested in determining when the evolution of whales from land-dwelling mammals occurred?

4. What role did the increase in oxygen concentration in the atmosphere play in the evolutionary history of life?

5. During what geologic period did most of the major living animal phyla appear in the fossil record?

6. During what geologic period did the supercontinent Pangaea begin to break up?

7. Approximately how many species are presently listed as threatened with extinction?

8. According to the fossil record, from which group of organisms did the mammals evolve?

A GUIDE TO THE READING

The following concepts typically give students the most difficulty when exploring the content in Chapter 20 for the first time. For each concept, one or more references have been identified that may help you gain a better understanding of these potentially problematic areas.

Carbon Dating

Radioisotopes, radioactive forms of chemical elements, occur naturally in the environment. These unstable substances undergo the process of radioactive decay, transforming into a more stable version of the element. The decay of the radioisotope is accompanied by the release of radioactive energy that we can measure as radioactivity. The key to understanding how this process can be used to determine the age of fossils is realizing that not only do these compounds exist naturally, but that the process of decay occurs naturally and at a constant rate as well. One radioactive substance used to date fossils is carbon-14. Recall from Chapter 5 that the element carbon typically has an atomic weight of 12. Carbon-14, in contrast, has an atomic weight of 14 because of the presence of additional neutrons, which make the nucleus unstable. When carbon-14 undergoes decay, the additional neutrons are released, yielding a stable carbon atom with an atomic weight of 12. The release of neutrons during the decay process is measured as radioactivity. The time it takes for half the amount of carbon-14 (for example, five of every 10 atoms) in a sample to decay to carbon-12 is 5,730 years. Therefore, by knowing that all fossils originated with roughly the same percentage of carbon-14 present, and by measuring the amount of carbon-14 that remains in a fossil, scientists are able to estimate the age of the fossil with relative accuracy.

For more information on this concept, be sure to focus on

• Section 20.1, The Fossil Record: A Guide to the Past
• Figure 20.1, Fossils through the Ages

Continental Drift

Although it may seem impossible, Earth's continents are in constant motion. The movement of the continents that occurs over time is called "continental drift" and results from the sliding of "plates" that float on the surface of the earth's liquid core. This is how large land masses, such as the Pangaea supercontinent that existed over 200 million years ago, have broken apart, their division resulting in the continents as we know them today. Recall from Chapter 19 that the geographic isolation of species that occurs when land masses separate in this fashion can contribute to speciation as populations become reproductively isolated and experience reduced gene flow.

For more information on this concept, be sure to focus on

- Section 20.3, The Effects of Plate Tectonics
- Figure 20.7, Movement of the Continents over Time

Adaptive Radiation

"Adaptive radiation" refers to the process that occurs when a group of organisms "expands" to inhabit new habitats and ecological roles. Adaptive radiation often occurs after various groups of organisms go extinct, removing ecological competition for resources. With reduced competition or predation, surviving organisms are more free to expand to take advantage of the resources available in an ecosystem. In other cases, adaptive radiation may occur after an organism has acquired an adaptation that allows it to survive or reproduce more efficiently within an ecosystem. Adaptive radiations can occur on a small or a large scale, depending on the supporting ecosystem.

For more information on this concept, be sure to focus on
- Section 20.5, Adaptive Radiations: Increases in the Diversity of Life

TYING IT ALL TOGETHER

Several concepts presented in this chapter build on those presented in previous chapters and may also be revisited and discussed in greater detail in subsequent chapters, including:

Classification of Life-forms on Earth

- Chapter 2—Section 2.2, The Linnaean System of Biological Classification

Origin of Life on Earth

- Chapter 2—Section 2.1, The Unity and Diversity of Life

Total Number of Species on Earth

- Chapter 2—Section 2.1, *The extent of Earth's biodiversity is unknown*

Oxygen and Metabolism

- Chapter 8—Section 8.2, Metabolism

Natural Selection

- Chapter 18—Section 18.6, Natural Selection: The Effects of Advantageous Alleles

Fossil Record

- Chapter 17—Section 17.4, *Evolution is strongly supported by the fossil record*

Adaptive Radiation of Galápagos Finches

- Chapter 17—Section 17.4, *Formation of new species can be observed in nature and can be produced experimentally*

Speciation and Geographic Isolation

- Chapter 19—Section 19.4, Speciation: Generating Biodiversity

PRACTICE QUESTIONS

Factual Knowledge

1. Which of the following continents was once a warm place covered with exotic plants and animals, but is now a cold desert where few organisms can survive?
   a. Antarctica
   b. North America
   c. Africa
   d. Pangaea
   e. Australia

2. Which of the following would *not* be considered an advantage for an organism that evolved the ability to walk upright?
   a. freeing hands to carry objects
   b. freeing hands to use tools
   c. improving line of sight
   d. greater mobility in trees
   e. none of the above

3. The first fossil evidence for life on Earth dates back to _____ years ago.
   a. 1.5 million
   b. 50 million
   c. 530 million
   d. 3.5 billion
   e. 4.6 billion

4. Which of the following is *not* a reason that the fossil record is incomplete?
   a. Decomposition often destroys organisms before they can fossilize.
   b. Evolution has not produced many organisms that fossilize well.
   c. The special circumstances needed for fossils to form are often not present.
   d. Geologic processes frequently destroy rocks before fossils are found.
   e. Locating fossils in rock formations is a difficult task.

5. Only rarely does the order in which organisms appear in the fossil record agree with our understanding of the evolutionary progression of life. (True or False)

6. Refer to Figure 20.1 in your textbook. Fossils such as those shown have been found in which of the following natural materials?
   a. sedimentary rock
   b. tree resin
   c. ice
   d. all of the above
   e. none of the above

7. Which of the following dates indicates the time when multicellular life first appeared on Earth?
   a. 4.6 billion years ago
   b. 3.5 billion years ago
   c. 650 million years ago
   d. 530 million years ago
   e. 65 million years ago

8. During the Cambrian explosion,
   a. a large and relatively sudden increase in the diversity of life took place.
   b. continental drift resulted in one of the five mass extinctions.
   c. there was no occurrence, according to the evidence, of adaptive radiation.
   d. humans first began driving organisms to extinction.
   e. none of the above

9. Of the following organisms, which one is believed to have been the first eukaryote to colonize the land?
   a. mushrooms
   b. green algae
   c. bacteria
   d. fish
   e. amphibians

10. Continental drift has played a major role in macroevolution, in part because the movement of continents helps create the geographic isolation that promotes speciation. (True or False)

11. Approximately when in Earth's history did the diversity of fish species suddenly increase?
   a. 65 million years ago
   b. 205 million years ago
   c. 250 million years ago
   d. 365 million years ago
   e. 445 million years ago

12. One way that mass extinctions affect the history of life is by
   a. eliminating entire groups of organisms.
   b. reducing ecological and evolutionary opportunities for other organisms.
   c. causing continents to move and climates to change.
   d. inhibiting adaptive radiation.
   e. all of the above

13. Which of the following statements about the evolutionary history of life on Earth is *false*?
   a. Mass extinctions have played a key role in macroevolution.
   b. Continental drift has shaped the course of evolution for many organisms.
   c. Evolution above the species level differs from evolution of populations.
   d. The overall diversity of life has decreased through time.
   e. none of the above

14. All species that have highly beneficial adaptations will survive a mass extinction. (True or False)

15. When hominids made the transition from life in the trees to life on the ground, which of the following also happened?
   a. Brain size decreased.
   b. Methods of locomotion began to change.
   c. The fully opposable big toe first appeared.
   d. Daytime vision was reduced.
   e. none of the above

16. The evolutionary origin of the primates dates back 65 million years to a group of nocturnal, tree-dwelling, insectivorous small mammals similar to modern tree shrews. (True or False)

17. Overall, the available evidence suggests that the best explanation for the origin of modern humans is the _____ hypothesis.
   a. multiregional
   b. out-of-Africa
   c. upright-posture
   d. trend-toward-big-brains
   e. chimpanzees-are-our-closest-relatives

18. Match each term with the best description.
    __ Cambrian explosion
    __ adaptive radiation
    __ fossils
    __ plate tectonics
    __ mass extinction
    a. process by which continents move over time
    b. process where groups of organisms expand to inhabit niches vacated by extinct groups
    c. period during which species numbers decline
    d. period during which species numbers increased dramatically
    e. preserved remains or imprints of organisms

Conceptual Understanding

1. The long-term, macroevolutionary changes in life that have occurred in Antarctica are best explained by
    a. continental drift.
    b. the Cambrian explosion.
    c. increases in atmospheric oxygen.
    d. fossilization.
    e. all of the above

2. Recent evidence suggests that modern humans are the direct descendants of Neandertals. (True or False)

3. A key factor that made eukaryotic life possible was
    a. continental drift.
    b. the Cambrian explosion.
    c. the colonization of land.
    d. reaching a threshold level of atmospheric oxygen.
    e. the elimination of bacterial competitors.

4. Which of the following macroevolutionary events occurred before the emergence of photosynthetic bacteria?
    a. Cambrian explosion
    b. first multicellular life
    c. first eukaryotes
    d. extinction of the dinosaurs
    e. none of the above

5. The emergence of humans, the extinction of the dinosaurs, and the colonization of land by plants and animals all
    a. were part of mass extinction events.
    b. occurred before the Cambrian explosion.
    c. are significant macroevolutionary events in the history of eukaryotes.
    d. took place before the emergence of multicellular life.
    e. none of the above

6. Fossilized human remains are often found in the same rocks that contain dinosaur fossils. (True or False)

7. Which of the following is *not* one of the challenges that animals had to face during the colonization of land?
    a. supporting the body
    b. reproduction
    c. conserving ions and water
    d. becoming multicellular
    e. movement

8. Of the following sequences, which one best represents the evolution of vertebrates?
    a. reptiles–amphibians–fish–birds–mammals
    b. amphibians–fish–reptiles–birds–mammals
    c. fish–reptiles–amphibians–mammals–birds
    d. fish–amphibians –reptiles–mammals–birds
    e. birds–mammals–reptiles–amphibians–fish

9. One of the major side effects of continental drift is _____, which in turn leads to _____.
    a. fossilization; adaptive radiation
    b. microevolution; climate change
    c. climate change; the extinction of many species
    d. mass extinction; a reduction in the amount of atmospheric oxygen
    e. release from competition; mass extinction

10. Of the five major mass extinctions, the one that had the most dramatic effect on macroevolution occurred
    a. 65 million years ago.
    b. 205 million years ago.
    c. 250 million years ago.
    d. 365 million years ago.
    e. 440 million years ago.

11. Adaptive radiations can be caused when species are released from competition. A dramatic example of this occurred 65 million years ago, when _____ emerged from the decline of _____.
    a. reptiles; amphibians
    b. mammals; dinosaurs
    c. insects; plants
    d. birds; mammals
    e. none of the above

12. The appearance of groups of animals in the fossil record does *not* match the order that is based on DNA and morphological evidence. (True or False)

13. Which of the following statements regarding the differences between mammalian and reptilian jaws and teeth is *false*?
    a. Extra muscles present in the cheek region allow for increased strength and precision of the mammalian jaw.
    b. The temporal fenestra present in modern mammals allows for the attachment of an extra muscle to the jaw.

c. The therapsids were the first group to develop a temporal fenestra.

d. Modern mammals have highly specialized teeth, which differ according to location within the jaw.

e. The enlarged jaw typical of mammalian species first appeared in the cynodonts.

14. Refer to Figure 20.8. The large increase in diversity that began 65 million years ago is mainly the result of
   a. adaptive radiation.
   b. continental drift.
   c. new fossil formation.
   d. high gene flow between populations.
   e. increased oxygen in the atmosphere.

15. A mass extinction has just occurred, wiping out 85 percent of all land animals. What are the key effects on the diversity of life after this mass extinction?
   a. One type of organism from all major groups will survive.
   b. New ecological and evolutionary opportunities will become available to those organisms that survive.
   c. Biodiversity will not change.
   d. all of the above
   e. none of the above

16. Suppose that a new fossil hominid is discovered. What criterion can be used to determine where in the human evolutionary lineage this new fossil belongs?
   a. brain size
   b. tooth structure
   c. foot structure
   d. evidence of tool use
   e. all of the above

RELATED ACTIVITIES

• Visit a few of the websites hosted by various natural history museums in the United States (for example, the Smithsonian, the Carnegie Museum of Natural History, the American Museum of Natural History, and the Field Museum of Chicago). Research the concept of macroevolution on these sites and compose a one-page summary describing how the exhibits at these locations illustrate the concept of macroevolution.

• Antarctica today is a biologically stark continent as opposed to how it was during the period when it was part of Pangaea and harbored abundant life. Consult a geology or biology reference source about the evolutionary history of Antarctica. Then create an evolutionary time line that shows how Antarctica drifted from a location nearer the equator to its current position over the South Pole. Be sure to include on your time line major evolutionary events of relevance to the organisms that lived on and around Antarctica.

• Use the Internet to research the discovery of an organism called "*Tiktaalik*," which represents a transitional form of life between ocean-dwelling fish and land animals. Using this discovery as a template, illustrate and describe in your own words what a missing link between land mammals and sea-dwelling mammals (for example, whales and dolphins) might have looked like.

ANSWERS AND EXPLANATIONS

Factual Knowledge

1. a. The macroevolutionary change that has occurred in Antarctica is stunning and an excellent example of the effect of continental drift. For more information, see the chapter Introduction, Puzzling Fossils in a Frozen Wasteland.

2. d. Recall that the ability to walk upright was an adaptation of hominids that enabled them to better adapt to life on land rather than in trees. For more information, see Section 20.7, *Walking upright was a big step in hominin evolution.*

3. d. The age of Earth is 4.6 billion years, but the first fossil life does not appear until 3.5–4 billion years ago. For more information, see Section 20.2, *The first single-celled organisms arose at least 3.5 billion years ago.*

4. b. Many of the organisms that have lived in the past could have potentially become fossils; however, because of the reasons given in answers a, c, d, and e, they did not produce fossils or the fossils were never found. For more information, see Section 20.1, *The fossil record is not complete.*

5. False. The two are often closely matched, providing strong support for the predictions made by evolutionary theory. For more information, see Section 20.1, The Fossil Record: A Guide to the Past.

6. d. Fossils can be found in a variety of materials, including those listed. For more information, see Section 20.1, The Fossil Record: A Guide to the Past.

7. c. There was a long period between the first appearance of life (3.5 billion years ago) and the evolution of multicellular organisms (650 million years ago). For more information, see Section 20.2, *Multicellular life evolved about 650 million years ago.*

8. a. Although the reasons for the Cambrian explosion remain unclear, it was one of the most dramatic examples of increased diversity in the history of life. For more information, see Section 20.2, *Multicellular life evolved about 650 million years ago.*

9. b. Bacteria were actually the first organisms to live on land, but they are prokaryotes. For more information,

see Section 20.2, *Colonization of land followed the Cambrian explosion.*

10. True. When continents separate, populations become isolated geographically. This reduces gene flow and promotes speciation. For more information, see Section 20.3, The Effects of Plate Tectonics.

11. e. In many ways, the increase in fish diversity 445 million years ago began the serious evolution of the vertebrate classes. For more information, see Figure 20.3, The Geologic Timescale and the Major Events in the History of Life.

12. a. The second and fourth answers tend to be increased by mass extinctions, whereas the third answer *causes* mass extinctions. For more information, see Section 20.4, Mass Extinction: Worldwide Losses of Species.

13. d. Overall diversity has increased, and dramatically so during the past 200 million years. See Figure 20.8, The Five Mass Extinctions Drastically Reduced the Diversity of Animals. For more information, see Section 20.4, Mass Extinction: Worldwide Losses of Species.

14. False. Survival of a mass extinction may be a purely random event, and having highly beneficial adaptations may not ensure survival. For more information, see Section 20.5, Adaptive Radiations: Increases in the Diversity of Life.

15. b. Life on the ground set the stage for upright walking, which quickly became the dominant mode of locomotion in hominids. For more information, see Section 20.7, *Walking upright was a big step in human evolution.*

16. True. Although the fossil record for early primates is sketchy, an animal that resembles the modern tree shrew is believed to be the mammal from which primates emerged 65–80 million years ago. For more information, see Section 20.7, *Walking upright was a big step in human evolution.*

17. b. The first answer is the only other serious choice, but the multiregional hypothesis has recently been called into question because it does not appear that the extensive gene flow required to make the hypothesis feasible would have been possible with widely spread populations of early *Homo.* For more information, see Section 20.7, *Modern humans spread out of Africa to populate the rest of the world.*

18. d. Cambrian explosion
b. adaptive radiation
e. fossils
a. plate tectonics
c. mass extinction
For more information, see Section 20.1, The Fossil Record: A Guide to the Past; Section 20.2, *Multicellular life evolved about 650 million years ago*; Section 20.4, Mass Extinctions: Worldwide Losses of Species;

and Section 20.5, Adaptive Radiations: Increases in the Diversity of Life.

Conceptual Understanding

1. a. On Antarctica, continental drift isolated all but the flying and swimming organisms. When drastic climate changes occurred, the majority of those organisms could not adapt and went extinct. For more information, see Section 20.3, The Effects of Plate Tectonics.

2. False. The Neandertals were an advanced type of archaic *Homo sapiens* that coexisted with humans in western Asia for some 80,000 years. For more information, see Section 20.7, *Modern humans spread out of Africa to populate the rest of the world.*

3. d. Eukaryotes are larger than prokaryotes, and so need more oxygen to survive. It wasn't until the proper atmospheric oxygen threshold was reached that eukaryotic life was possible. For more information, see Section 20.2, *The first single-celled organisms arose at least 3.5 billion years ago.*

4. e. The evolution of photosynthetic bacteria occurred very early in the history of life, predating all of the events listed. See the time line in Figure 20.3. For more information, see Section 20.2, *The first single-celled organisms arose at least 3.5 billion years ago.*

5. c. Recall that macroevolution focuses on the overall pattern of change over time. These events are all significant macroevolutionary events that have occurred in the history of eukaryotes on earth. For more information, see Section 20.2, The History of Life on Earth.

6. False. Humans and dinosaurs are separated on the geologic timescale by over 50 million years of evolution. For more information, see Section 20.2, The History of Life on Earth.

7. d. The first land animals would have already been multicellular organisms. For more information, see Section 20.2, *Colonization of land followed the Cambrian explosion.*

8. d. Vertebrate life first began in water with fish, then partially moved to land with amphibians, and subsequently became specialized for the terrestrial world with reptiles, mammals, and birds. Refer to the geologic timescale in Figure 20.3, The Geologic Timescale and the Major Events in the History of Life. For more information, see Section 20.2, The History of Life on Earth.

9. c. As continents move, climates change because of distance from the equator and shifts in ocean currents. These changes can force species into extinction. For more information, see Section 20.3, The Effects of Plate Tectonics.

10. c. The great Permian mass extinction saw enormous numbers of species and other taxonomic groups completely disappear, especially from the marine environment. For more information, see Section 20.4, Mass Extinctions: Worldwide Losses of Species.

11. b. The mass extinction of 65 million years ago eliminated the dinosaurs and removed a major source of competition for mammals, whose numbers and varieties then increased. For more information, see Section 20.5, Adaptive Radiations: Increases in the Diversity of Life.

12. False. The fossil record agrees very well with the genetic and morphological evidence to provide support for evolution. For more information, see Section 20.1, The Fossil Record: A Guide to the Past.

13. c. Recall that the temporal fenestra was an early adaptation that first appeared in ancestral reptiles (*Haptodus*). Review Figure 20.12, From Reptile to Mammal for additional information. For more information, see Section 20.6, *The mammalian jaw and teeth evolved from reptilian forms in three stages.*

14. a. The increase is an adaptive radiation caused by the removal of many species, including the dinosaurs, through a mass extinction event 65 million years ago. For more information, see Section 20.5, Adaptive Radiations: Increases in the Diversity of Life.

15. b. When mass extinctions occur, entire groups of organisms are lost. This tends to open up more ecological and evolutionary opportunities to those organisms that do survive. For more information, see Section 20.5, Adaptive Radiations: Increases in the Diversity of Life.

16. e. All of these hominid characteristics have been influenced by evolution, and so they serve as legitimate bases for comparison of the new fossil to existing fossils. For more information, see Section 20.7, Human Evolution.

CHAPTER 21 | The Biosphere

GETTING STARTED

Below are a few questions to consider before reading Chapter 21. These questions will help guide your exploration and assist you in identifying some of the key concepts presented in this chapter.

1. What is the relationship between the zebra mussel and cargo ship ballast water?

2. What is the estimated annual impact on the U.S. economy that is attributable to introduced species?

3. Why are Earth's equatorial regions so much warmer than its polar regions?

4. Why are most deserts located near 30 degrees latitude?

5. What best explains why water temperatures on the Pacific coast of the United States are so much cooler than on the Atlantic coast?

6. What two climatic factors most strongly influence the development of a terrestrial biome?

7. Terrestrial biomes are named after the dominant vegetation of the region. How are aquatic biomes named?

8. What two factors most strongly influence the development of an aquatic biome?

A GUIDE TO THE READING

The following concepts typically give students the most difficulty when exploring the content in Chapter 21 for the first time. For each concept, one or more references have been identified that may help you gain a better understanding of these potentially problematic areas.

The Importance of Ecology

Chapter 21 begins with the question "Why is ecology important?" Although a complete answer is beyond our current understanding, it has become increasingly evident that people are dependent on the continuous operation of the biosphere's natural systems. Unfortunately, human activity is changing the biosphere in many ways, some of which may not be at all positive. Introduced species clearly illustrate this concern. Whether introduced intentionally or accidentally, plant and animal species from other regions may expand uncontrollably, displace native species, and reduce the usefulness or dependability of natural systems. Current estimates for economic damage related to introduced species are near $120 billion annually for the United States alone. Once it is established, removing an introduced species becomes all but impossible. Yet banning all future introductions may not be in our best interests either. More than 98 percent of the food currently produced in the United States comes from introduced crops. Plant species with potentially valuable uses are routinely discovered, and, despite our best efforts, accidental introductions will continue. A more complete understanding of ecology can help us understand the consequences of these and other environmentally damaging events (such as the dispersal of toxic chemicals), direct an appropriate response to current problems, and perhaps prevent future ones from occurring.

For more information on this concept, be sure to focus on

- Section 21.1, Ecology: Understanding the Interconnected Web

Climate Has a Large Effect on the Biosphere

Earth's surface is not uniformly heated. Sunlight strikes the equator and Earth's tropical regions directly, but it strikes the poles and higher latitudes at a slant. Although these angles change somewhat with the seasons, the tropics experience the least variation, providing a relatively warm, stable climate throughout the year for the organisms living there. Polar regions cycle through relatively short, mild summers to the bitter cold of winter. Such extremes over the course of a year make life in the polar regions much more challenging than life in the tropics. Differences in the temperature of Earth's surface heat or cool the adjacent atmosphere. In tropical regions, the moist air warms and rises, but it expands and cools when doing so. Because cool air has a lower capacity to hold moisture, rain will eventually fall. In addition, the loss of moisture has also reduced the air's weight, and so it is unable to sink through the moist air rising from the surface beneath it. The air is first displaced to the north and south, eventually sinking at about 30 degrees latitude. As it descends the air is compressed, causing it to warm and gain moisture from Earth's surface as it moves back toward the equator. By the time it reaches the equator, the air is warm and moist, and the cycle is repeated. This cyclical circulation of air is termed a "convection cell." Two convection cells exist in each hemisphere: (1) a polar cell, with air descending at the pole and flowing to about 60 degrees latitude, where it rises, and (2) the tropical cell just described. The horizontal surface component of each cell produces the winds that most directly influence the biosphere. These winds are relatively consistent in the tropics and polar regions but somewhat variable in the temperate regions between 30 degrees and 60 degrees latitude. The winds, however, do not move directly north or directly south, as the orientation of the convection cells would suggest. Influenced by the rotation of Earth, they curve to the right in the northern hemisphere and to the left in the southern. In the northern hemisphere the surface winds blowing south toward the equator turn to the right, appearing to come from the east; they are called "easterlies." Surface winds traveling poleward are also turned to the right, appear to come from the west, and are called "westerlies."

For more information on this concept, be sure to focus on

- Section 21.2, *Incoming solar radiation shapes climate*
- Section 21.2, *Wind and water currents affect climate*
- Figure 21.2, Earth Has Four Giant Convection Cells
- Figure 21.3, Prevailing Winds Are Determined by Global Patterns of Air Circulation

Biomes

The biosphere can be divided into several terrestrial and aquatic life zones, termed "biomes." Seven terrestrial and eight aquatic biomes are commonly recognized. Terrestrial biomes are named for the dominant forms of vegetation within each region. Examples include forests and grasslands. Rainfall and temperature are the primary factors that influence biome development. When rainfall is abundant, a forest can form. Recall that rainfall is high at the equator and at 60

degrees latitude. Forests exist in both locations, but the specific form of forest—a tropical rainforest or boreal forest, for example—is determined primarily by average temperature. Because water is, thermally, relatively stable, aquatic biomes are less influenced by temperature than are terrestrial biomes. Thus the annual temperature range for most aquatic biomes is small. Climate, however, still significantly influences most aquatic biomes. As was true of air, the temperature of water directly affects its density. Vertical mixing can occur in lakes and the ocean when surface waters cool and sink. In lakes, such mixing may move oxygen-rich surface water to the bottom and displace nutrient-rich water toward the surface. Marine aquatic biomes may be affected by temperature in additional ways. High temperatures may evaporate water from a marine biome, concentrating salt and increasing salinity, whereas rainfall may dilute salt water and lower salinity. Although the situations just described tend to be localized, physical conditions may sometimes alter a significant portion of an entire ocean basin. Consider the El Niño events that develop periodically off the western coast of South America. Unusually warm ocean water floats over the top of cooler nutrient-rich water, preventing it from reaching the surface. Without the replenishment of nutrients, the abundance of marine life declines spectacularly.

For more information on this concept, be sure to focus on

- Section 21.3, Terrestrial Biomes
- Section 21.4, *Aquatic biomes are influenced by terrestrial biomes and climate*
- Section 21.4, *Aquatic biomes are also influenced by human activity*
- Figure 21.15, El Niño Events

TYING IT ALL TOGETHER

Several concepts presented in this chapter build on those presented in previous chapters and may also be revisited and discussed in greater detail in subsequent chapters, including

The Importance of Ecology

- Chapter 22—Section 22.4, Logistic Growth and the Limits on Population Size

Interactions with the Environment

- Chapter 23—Section 23.3, *Succession establishes new communities and replaces disturbed communities*

Climate Has a Large Effect on the Biosphere

- Chapter 23—Section 23.3, *Communities change as climate changes*

Biomes

- Chapter 24—Section 24.2, *The rate of energy capture varies across the globe*

PRACTICE QUESTIONS

Factual Knowledge

1. One major difference between weather and climate is that
   a. weather changes rapidly but climate is predictable.
   b. climate shifts rapidly but weather is predictable.
   c. weather is cyclical.
   d. climate is the weather during a short interval.
   e. none of the above

2. Seasons are influenced by
   a. the tilt of Earth on its axis.
   b. the amount of solar radiation reaching Earth's surface.
   c. Earth's movement around the sun.
   d. all of the above
   e. none of the above

3. The term "convection cell" refers to
   a. the rising of cool, dry air and the descending of warm, moist air.
   b. a factor that contributes to wind patterns.
   c. the rising of warm, moist air and the descending of cool, dry air.
   d. both a and b
   e. both b and c

4. Deserts are most likely to occur
   a. at 30 degrees above and below the equator.
   b. in the rain shadows of mountains.
   c. where cool, dry air descends.
   d. all of the above
   e. none of the above

5. Places like London, England, are generally warmer than places of the same latitude in North America because
   a. the English Channel acts as insulation.
   b. the Gulf Stream transfers heat from the tropics.
   c. sunlight strikes the eastern hemisphere more directly.
   d. all of the above
   e. none of the above

6. Biomes are generally uniform in their species distribution. (True or False)

7. The makeup of any given biome is affected by which of the following factors?
   a. climate
   b. the species residing there
   c. geography
   d. human impact
   e. all of the above

8. Direct exclusion of a species from a biome means that
   a. the species cannot survive the climate of the biome.
   b. the species can survive the climate but is outcompeted by species better suited to the climate.
   c. human intervention has prevented the species from establishing itself.
   d. all of the above
   e. none of the above

9. In regions of hot temperatures and wet climate, you will most likely find _____ biomes, whereas in regions of hot temperatures and dry climate, you will find _____ biomes.
   a. desert; tropical
   b. temperate; arid
   c. tropical; desert
   d. arid; temperate
   e. tundra; chaparral

10. Terrestrial biomes are classified according to which of these characteristics?
    a. physical characteristics of the environment
    b. dominant vegetation living there
    c. human impact on the biome
    d. all of the above
    e. none of the above

11. Aquatic biomes of the same type are very similar in terms of the species distributed within them. (True or False)

12. Terrestrial and aquatic biomes have little effect on each other. (True or False)

13. El Niño affects the coast of Peru by
    a. changing the flow of the current.
    b. warming the waters off the coast.
    c. changing the distribution of species off the coast.
    d. influencing weather patterns.
    e. all of the above

14. Match each term with the best description.
    ___ biosphere
    ___ ecology
    ___ biome
    a. the study of interactions between organisms and their environment

b. all of the organisms and the environments in which they live

c. one of several major terrestrial or aquatic life zones

15. The hole created in the ozone layer by chlorofluorocarbons is problematic because
    a. heat from Earth is escaping through the hole.
    b. ozone protects us from the sun's ultraviolet light.
    c. it leads to global warming.
    d. all of the above
    e. none of the above

Conceptual Understanding

1. Why is the study of ecology becoming increasingly important?
   a. Humans are increasingly affecting the environment.
   b. We truly understand very little about how organisms and their environments interact.
   c. Human population growth is putting more and more biomes at risk.
   d. all of the above
   e. none of the above

2. Occasionally the Gulf Stream follows a more southerly path. How might this affect the climate of Northern Europe?
   a. produce greater temperature ranges, with hotter summers and colder winters
   b. produce reduced temperature ranges, with warmer winters and cooler summers
   c. produce a general cooling, resulting in cooler winters and summers
   d. produce a general warming, resulting in warmer winters and summers
   e. The Gulf Stream has no appreciable effects on the climate of Northern Europe.

3. You have been teleported to a terrestrial region of Earth you have never visited before. You determine which biome you are in by noting the
   a. climate.
   b. dominant vegetation living there.
   c. most obscure species you can find.
   d. both a and b
   e. both a and c

4. Climate events like El Niño are caused by changes in ocean currents. This indicates that ocean currents have a _____ impact on _____ climate.
   a. small; regional
   b. small; global
   c. large; global

   d. large; regional
   e. none of the above

5. You are in an area in which the water fluctuates between very salty and not so salty. You look around and see that the region is located at the place where a river meets the ocean. You are most likely in a(n) _____ biome.
   a. estuarine aquatic
   b. river aquatic
   c. tundra terrestrial
   d. intertidal aquatic
   e. none of the above

6. A friend writes to tell you that his community is very cold and there are many days of total darkness. They haven't had snow in quite some time, but the permafrost layer is very deep. Your friend is writing to you from the
   a. chaparral.
   b. boreal forest.
   c. tundra.
   d. desert.
   e. none of the above

7. Overfishing is one way humans affect aquatic biomes. The most serious aspect of overfishing is
   a. the oil and gas pollution associated with boat fuel tanks.
   b. the introduction of live bait into the biome.
   c. its potential to change the species distribution within the biome.
   d. noise pollution from boat motors.
   e. none of the above

8. The eastern side of the Rockies tends to receive much less rainfall than the western side because of
   a. convection currents.
   b. the rain shadow effect.
   c. human impact.
   d. all of the above
   e. none of the above

9. Equatorial regions are tropical because
   a. the sun strikes them more directly than it does other parts of Earth.
   b. hot moist air rises, then cools and causes rain.
   c. large bodies of water provide ample moisture.
   d. all of the above
   e. none of the above

10. The biosphere depends on
    a. interactions among all organisms and the environment.
    b. humans living responsibly within it.
    c. physical factors such as climate.

d. all of the above
e. none of the above

RELATED ACTIVITIES

- Many biological supply houses sell what they call "mini-biospheres," enclosed habitats that sustain a number of different species indefinitely. Using the Internet or a supply house catalog, find out more about these minibiospheres. Compare and contrast them to the biosphere.
- Investigate how human impact has changed the distribution of biomes over the past 100 years. Determine if there is anything we can do to reverse these effects.
- Select a biome of interest. Discuss how variable that biome is with respect to species distribution. Discuss which factors affect its species distribution.
- Investigate the greenhouse effect. What are its causes and what do scientists predict its consequences will be?

ANSWERS AND EXPLANATIONS

Factual Knowledge

1. a. Weather changes daily, if not more often. Climate follows patterns over long periods. For more information, see Section 21.2, Climate's Large Effect on the Biosphere.
2. d. Seasons are influenced by the amount of solar radiation reaching a region at a particular time, which in turn is influenced by the tilt of Earth and the movement of Earth around the sun. For more information, see Section 21.2, *Incoming solar radiation shapes climate*.
3. e. Warm, moist air rises and cools; dry air descends. This movement, coupled with the rotation of Earth on its axis, contributes to wind patterns. For more information, see Section 21.2, *Wind and water currents affect climate*.
4. d. All of the listed factors contribute to desert formation, as does human impact. For more information, see Section 21.3, *The scarcity of moisture shapes life in the desert*.
5. b. The Gulf Stream acts as a major warming force for western Europe. For more information, see Section 21.2, *Wind and water currents affect climate*.
6. False. Even though biomes are characterized by a dominant species, the distribution of that species varies. For more information, see Section 21.3, *The location of terrestrial biomes is determined by climate and human actions*.
7. e. Biomes are dynamic entities and are influenced by multiple factors. For more information, see Section

21.3, *The location of terrestrial biomes is determined by climate and human actions*.

8. a. Climate is responsible for direct exclusion of a species from a particular biome. Species may be indirectly excluded from a biome by other species. For more information, see Section 21.2, *Wind and water currents affect climate*.
9. c. Temperature and precipitation affect the distribution of biomes. Section 21.3, *Tropical forests have high species diversity* and *The scarcity of moisture shapes life in the desert*.
10. b. Terrestrial biomes are classified by their dominant life-forms. For more information, see Section 21.3, *The location of terrestrial biomes is determined by climate and human actions*.
11. False. Though physical characteristics may be similar, species distribution may be quite different. For more information, see Section 21.4, *Aquatic biomes are influenced by terrestrial biomes and climate*.
12. False. Terrestrial and aquatic biomes have remarkable influences on one another. For more information, see Section 21.4, *Aquatic biomes are influenced by terrestrial biomes and climate*.
13. e. The changing currents of El Niño have large and widespread effects. For more information, see Section 21.4, *Aquatic biomes are influenced by terrestrial biomes and climate*.
14. b. biosphere. For more information, see the chapter Introduction, A View of Earth from Space.
    a. ecology. For more information, see the chapter Introduction, A View of Earth from Space.
    c. biome. For more information, see Section 21.3, Terrestrial Biomes.
15. b. The ozone layer prevents much of the sun's ultraviolet light from reaching Earth and thereby protects organisms from mutations. See the chapter's Biology Matters box, "Wearing Thin: The Attack on Earth's Ozone Shield."

Conceptual Understanding

1. d. Many questions have yet to be answered about the environment, its organisms, and the impact of humans on both. For more information, see Section 21.1, Ecology: Understanding the Interconnected Web.
2. c. The Gulf Stream transfers a tremendous amount of heat energy that moderates the climate of northern Europe; without this influence both summers and winters would be cooler. For more information, see Section 21.2, *Wind and water currents affect climate*.
3. b. Terrestrial biomes are categorized according to the dominant vegetation living there. For more information, see Section 21.3, Terrestrial Biomes.

4. c. El Niño has caused major global events. For more information, see Section 21.4, *Aquatic biomes are influenced by terrestrial biomes and climate.*

5. a. An estuarine environment occurs where salt water and fresh water mingle. For more information, see Section 21.4, *Estuaries and coastal regions are highly productive parts of the marine biome.*

6. c. These are all characteristics of tundra. For more information, see Section 21.3, *A few coniferous species dominate in the boreal forest.*

7. c. The depletion of one or more species by overfishing can dramatically alter the ecological balance among remaining species. For more information, see Section 21.4, *Aquatic biomes are also influenced by human activity.*

8. b. Mountains are large geographic barriers that greatly influence climate. For example, mountains may create a rain shadow in which little precipitation falls on the side opposite the prevailing winds. For more information, see Section 21.2, *The major features of Earth's surface also shape climate.*

9. d. Regions at the equator are tropical because sunlight strikes them directly, and as the hot, moist air rises, it cools and causes rain. Large bodies of warm water also influence the climate. Tropical regions are some of the most productive and diverse regions in the biosphere. For more information, see Section 21.3, *Tropical forests have high species diversity.*

10. d. Humans must live responsibly to limit their impact on the biosphere. We have the ability to affect nearly all the factors that influence the biosphere, including climate and the interactions between other organisms and the environment. For more information, see Section 21.1, Ecology: Understanding the Interconnected Web.

CHAPTER 22 | Growth of Populations

GETTING STARTED

Below are a few questions to consider before reading Chapter 22. These questions will help guide your exploration and assist you in identifying some of the key concepts presented in this chapter.

1. How many days would be required for an *E. coli* population, growing exponentially, to produce as many individuals as there are atoms in the universe?

2. In what circumstances can exponential growth be observed?

3. What pattern of growth is seen when essential resources become limited?

4. What happened to the reindeer population of Saint Paul Island?

5. How are the population densities of the snowshoe hare and the Canada lynx related?

6. What types of information would biologists need to be able to predict the impact of logging on spotted owl populations?

7. How many humans can Earth support?

8. In what ways have humans altered their carrying capacity during the past 10,000 years?

9. How many acres are required to support the average citizen of the United States?

A GUIDE TO THE READING

The following concepts typically give students the most difficulty when exploring the content in Chapter 22 for the first time. For each concept, one or more references have been identified that may help you gain a better understanding of these potentially problematic areas.

What Are Populations?

A population is a group of individuals of the same species living and interacting in a common geographic location. As a result of additions to the population through births and immigration and losses from death and emigration, population size has the potential to change with time. Biologists often find it convenient to focus on the biological factors affecting populations. Although the movements of individuals may be locally important, they have no net effect on the total number of individuals; one individual joining a population simply means that it has left another.

For more information on this concept, be sure to focus on

- Section 22.1, What Is a Population?
- Section 22.2, Changes in Population Size

Patterns of Change

The keys to understanding how populations change are recognizing the importance of resources, understanding that they are limited in availability, and realizing that the individuals of any population require many different resources, each being more or less common. Even when food is abundant, for example, if the appropriate cover required by juveniles is limited, most will be found and taken by predators, and the population will grow slowly or not at all. Populations display a variety of growth patterns in nature, but two patterns are observed repeatedly. Both are best understood as they relate to resource availability. During exponential growth, a constant proportion of individuals is added during each unit of time. Such growth occurs only when the population is too small to have an appreciable impact on the quantity of resources provided by the environment. When represented graphically, the number of individuals plotted against time produces a curve that resembles the letter "J." Most populations, however, do not maintain a consistent growth rate over time. As resources become limited, the growth rate declines. This type of population growth can be represented by an S-shaped curve. Such populations may appear to be growing exponentially at first, but eventually they stabilize. Biologists use the term "carrying capacity" to describe this level. It is important to understand that these patterns of growth do not represent an either-or situation. Many populations can grow exponentially when their numbers are small, but later they convert to an S-shaped pattern as resources become less common. Population growth models can be easily constructed using spreadsheets (see "Related Activities" later in this chapter), but such representations are often oversimplified. A constant population size implies a constant replenishment of resources.

A good example is the regrowth of prairie grasses during the spring from their dormant root systems. Recall, however, that organisms can substantially influence the physical environment. For example, too many hooves seeking a limited amount of grass can result in soil erosion, fewer surviving plants, and less grass for grazing at a later time. Most populations experience variations in carrying capacity, resulting either from direct influences from the population itself or, more commonly, from climatic factors. In the last chapter we read about the El Niño event. You probably realize now that it's the carrying capacity of the open ocean biome that changes when ocean circulation patterns change. When population densities are high, virtually all interactions between population members become more stressful. Competition for food, space, and shelter from predators and weather intensifies. Less food, or sunlight, makes an individual more susceptible to disease, and the high number of individuals improves the chances that a disease will quickly spread within the population. When the impact of any factor causes the density of the population to change, that factor can be described as being density-dependent. Alternatively, some factors are considered density-independent. One way to distinguish between density-dependence and density-independence is to consider the likelihood of an event relative to population size. Consider predation—when competition for food is low, prey animals can feed in relative safety. When population densities are high, the safe food is quickly eaten, leaving only the more exposed food sources available. Venturing out into the open to nibble a blade of grass may be an animal's last meal.

For more information on this concept, be sure to focus on

- Section 22.3, Exponential Growth
- Section 22.4, *Growth is limited by essential resources and other environmental factors*

- Section 22.4, *Some growth-limiting factors depend on population density; others do not*
- Figure 22.12, Same Species, Different Outcomes

The Human Population

It is no surprise that one of the most important issues facing population ecologists is the future of human growth. Understanding population growth has provided ecologists with some important predictive abilities, and many wonder if these are applicable to the human population. The key to understanding the connection to human growth, and the principles discussed in this chapter, is recognizing that the human carrying capacity has not been static. Technological advances have repeatedly increased the human carrying capacity. Advances in medicine have extended life expectancy and lowered infant mortality. Advances in agriculture have stabilized the food supply in many developed regions. Yet while the human growth curve continues to maintain its J-shape, poverty, starvation, disease, and other indicators of density dependence are becoming all too common. Will new technologies extend the carrying capacity even further, or do the declining lifestyles of so many of the world's citizens represent the future of humanity?

For more information on this concept, be sure to focus on

- Section 22.5, Patterns of Population Growth
- Figure 22.13, Rapid Growth of the Human Population

TYING IT ALL TOGETHER

Several concepts presented in this chapter build on those presented in previous chapters and may also be revisited and discussed in greater detail in subsequent chapters, including

What Are Populations?

- Chapter 25—Section 25.1, Land and Water Transformation

Patterns of Change

- Chapter 23—Section 23.1, *In exploitation, one member benefits while another is harmed*
- Chapter 23—Section 23.2, *In competition, both species are negatively affected*
- Chapter 23—Section 23.3, How Communities Change over Time

The Human Population

- Chapter 25—Biology Matters, "Toward a Sustainable Society"

PRACTICE QUESTIONS

Factual Knowledge

1. All populations
   a. eventually crash and disappear as a result of environmental deterioration.
   b. contain individuals of one species that interact in the same location.
   c. change their size only in response to the birth of new individuals.
   d. grow exponentially.
   e. all of the above

2. Today, on average, 31 people per square kilometer live in the United States. This figure is an example of a(n)
   a. carrying capacity.
   b. environmental limit.
   c. population density.
   d. density-dependent limit on growth.
   e. density-independent limit on growth.

3. Two factors that contribute to a decline in population size are _____ and _____.
   a. birth; death
   b. birth; immigration
   c. death; immigration
   d. death; emigration
   e. none of the above

4. In exponential growth, the
   a. numerical increase in the population remains constant for each generation.
   b. population size is usually above the environmental carrying capacity.
   c. population increases by some constant proportion of its current size.
   d. all of the above
   e. none of the above

5. Which of the following can be a limit on population growth?
   a. food
   b. water
   c. space
   d. accumulated wastes
   e. all of the above

6. Refer to Figure 22.6 in your textbook, which shows the growth of a population of *Paramecium caudatum* in a laboratory culture. The straight line of the growth curve, corresponding to the carrying capacity for this population, is the level at which
   a. the number of births equals the number of deaths.
   b. nutrients are extremely abundant.
   c. excess space is abundant.
   d. accumulated wastes are low.
   e. all of the above

7. Which of the following is likely to cause an increase in human population size?
   a. advancements in agriculture
   b. better weather predictions
   c. reducing the number of highway deaths
   d. improvements in medicine
   e. all of the above

8. Volcanic eruptions represent an example of density-dependent limits on population growth. (True or False)

9. The graph in Figure 22.8 is an example of a
   a. situation in which food and space were not limiting for the population.
   b. population that exceeded its carrying capacity and then crashed.
   c. density-independent limit on population growth.
   d. population cycle.
   e. all of the above

10. Which of the following statements about population growth is *false*?
    a. Tightly linked changes between populations of two species are common.
    b. Different populations of the same species may experience different growth patterns.
    c. Exponential population growth cannot continue indefinitely.
    d. One problem with large populations is that they cause habitat deterioration.
    e. Food shortages, predators, and weather can limit population growth.

11. The spotted owl is an example of a species in which population growth
    a. is declining because of natural factors such as fire and floods.
    b. is currently exponential and numbers are increasing rapidly.
    c. depends on the amount and arrangement of preferred habitat.
    d. has been stable for almost 100 years.
    e. resembles the growth of the human population on Easter Island.

12. If human birth rates were to drop immediately to a level that would *not* result in a population increase, the population would continue to grow for decades. (True or False)

13. One reason human populations have undergone such tremendous growth is that we have
    a. colonized a wide range of new habitats.
    b. greatly increased the carrying capacity of the places in which we live.

c. eliminated predators and many kinds of disease from our lives.

d. created a situation in which our birth rate exceeds our death rate.

e. all of the above

14. "Sustainable use of resources" refers to use that will *not* deplete critical resources and cause serious damage to the environment. (True or False)

15. Match each term with the best description.
    ___ carrying capacity
    ___ population size
    ___ exponential growth
    ___ S-shaped growth curve
    ___ population density
    a. one form of rapid population increase
    b. shows common form of population growth
    c. maximum sustainable population size
    d. total number of individuals in a population
    e. number of individuals per unit area

Conceptual Understanding

1. Which of the following examples of population growth in nonhuman organisms shows a pattern similar to that of the humans on Easter Island?
   a. Opuntia cactus in Australia
   b. reindeer on Saint Paul Island in Alaska
   c. paramecia in Gause's laboratory
   d. snowshoe hares and lynx in Canada
   e. spotted owls in the U.S. Pacific Northwest

2. Natural populations are easy to define. (True or False)

3. The growth of Opuntia cactus in Australia prior to the introduction of the Cactoblastis moth was
   a. slow.
   b. moderate.
   c. exponential.
   d. density-dependent.
   e. limited by shortages of nutrients and space.

4. Closed laboratory systems differ from natural systems in having
   a. no immigration or emigration.
   b. limited nutrients.
   c. limited space.
   d. all of the above
   e. none of the above

5. If birth and immigration levels are lower than death and emigration levels, the number of individuals in a population of organisms will
   a. grow.
   b. decline.
   c. stay the same.

d. double.
e. none of the above

6. The population growth patterns of giant puffball mushrooms, prickly pear cacti, aphids, and reindeer all demonstrate that
   a. limits exist, and no population can increase in size indefinitely.
   b. every population shows an S-shaped growth curve.
   c. limiting factors on population growth are all density-dependent.
   d. limiting factors on population growth are all density-independent.
   e. natural populations are not sustainable.

7. Many scientists now believe that the human population is at or near its carrying capacity on this planet. (True or False)

8. When carrying capacity drops, the number of individuals that can be supported in a habitat increases. (True or False)

9. Which of the following activities affecting human population growth is considered sustainable?
   a. the use of oil and gas as fuels for heating and transportation
   b. having family sizes greater than two children per female parent
   c. relying more heavily on irrigated crop lands for food production
   d. altering the amount of biodiversity through deforestation
   e. none of the above

10. Even though Americans represent a relatively small percentage of the world's human population, we have a disproportionately large impact on human population growth because we
    a. consume a very high percentage of the world's natural resources.
    b. cannot grow enough food to support ourselves.
    c. don't contribute much to advancements in health and medicine.
    d. have families with very large numbers of children.
    e. have a population doubling time much greater than that of most other countries.

11. If human populations continue to grow at their current rate, by the year 2025 there could be
    a. over 1 billion people on Earth.
    b. over 8 billion people on Earth.
    c. over 1 billion people in the United States.
    d. over 8 billion people in the United States.
    e. fewer people on Earth than there are today.

12. An examination of the numbers of bald eagles in the United States over the past 30 years reveals that
    a. decreasing numbers of prey species prompted declines.
    b. populations can recover from the impact of humans.
    c. species on the road to extinction continue to decline in numbers.
    d. all of the above
    e. none of the above

RELATED ACTIVITIES

• Consider the 1993 hantavirus outbreak in the American Southwest. This is a good example of how a deadly human disease is affected by population growth in non-human organisms (in this case, deer mice). Use your library or the Internet to research human diseases that are spread by animals such as mosquitoes, ticks, and rodents. Then write a short essay on the similarities and differences between these diseases. In particular, focus on how the population growth of the animals carrying the diseases influences the likelihood of human infection.

• Study the website maintained by the organization Zero Population Growth (www.zpg.org) and write a short essay describing the organization's goals and activities. Include your own opinion of the political and social agendas of this organization.

• Write a letter to the editor of your local newspaper on the issue of human population growth. Briefly describe the biology behind the problem, as well as your own personal views about how we can minimize the potential for ecological disaster. Be sure to include suggestions for ways we can help reduce the problem of overpopulation.

• Use an Excel spreadsheet to examine J-shaped and S-shaped growth. Create a time series in Column A. Enter 1 in cell A1. Enter this formula in cell A2 = (A1+1), and use the fill handle to fill down to A35. Enter identical initial population sizes (20) in cells B1 and C1, an initial *lambda* (1.25) in cell D1, and a carrying capacity value in cell E1 (350). To model J-shaped growth, enter this formula in cell B2 = B1 + B1*(C1 – 1), and use the fill handle to fill down through B35. To model S-shaped growth, enter this formula in cell C2 = C1 + C1*((D1– 1)*(E1–C1)/E1), and fill down through C35. Select the block A1:C35 and use the Chart Wizard to construct a Scatter Graph with smoothed lines. To see S-shaped growth on the graph, you will need to adjust the maximum value for the Y axis. Simply change the initial population size, *lambda*, or carrying capacity to explore the impact these values have on the growth curves. Write

a brief summary describing the difference between the growth models.

• Visit the Global Footprint Network web page mentioned in the chapter's Biology Matters box, "How Big Is Your Ecological Footprint?" Determine your ecological footprint by using the footprint calculator. Why is your footprint the size it is? What can *you* do to change your number? Make a list of your changes and how you believe they can decrease the size of your footprint.

ANSWERS AND EXPLANATIONS

Factual Knowledge

1. b. This is the basic definition of any population. For more information, see Section 22.1, What Is a Population?

2. c. Population measurements that indicate the number of individuals per unit area are known as "densities." For more information, see Section 22.1, What Is a Population?

3. d. Death and emigration are ways that individuals leave a population, and so they contribute to a decline in numbers. For more information, see Section 22.2, Changes in Population Size.

4. c. Exponential growth occurs when the population increases by some proportional amount (doubling, for example) in each successive generation. Such growth produces a very rapid increase in numbers. For more information, see Section 22.3, Exponential Growth.

5. e. The first three factors represent important resources that populations need in order to reproduce. The fourth is a potential negative factor for survival and reproduction. Because reproduction is an important contributor to population growth, all four factors can limit growth. For more information, see Section 22.4, *Growth is limited by essential resources and other environmental factors*.

6. a. At this point there is no net population growth, and so by definition the number of births equals the number of deaths. The other three conditions are more likely to occur when the population size is well below the carrying capacity of the environment. For more information, see Section 22.4, *Logistic population growth is the norm in the real world*.

7. e. All of these improvements in human life would likely increase survival and reproduction and therefore lead to increases in population size. For more information, see Section 22.4, *Logistic population growth is the norm in the real world*.

8. False. Volcanic eruptions fall into the same category as fires, floods, and weather variations. Such phenom-

ena are density-independent limits on growth. For more information, see Section 22.4, *Logistic population growth is the norm in the real world.*

9. b. Clearly, the population of reindeer grew well initially, as evidenced by the exponential rise in numbers during their first 20 years on Saint Paul Island. However, at its peak the population exceeded the carrying capacity of the environment, and then it crashed to extinction before the environment had a chance to recover enough to sustain the population. For more information, see Section 22.4, *Logistic population growth is the norm in the real world.*

10. a. Long-term, tightly linked changes between two populations, as seen with the snowshoe hares and the lynx, are actually rather uncommon in nature. For more information, see Section 22.5, Patterns of Population Growth; Section 22.3, Exponential Growth; and Section 22.4, *Some growth-limiting factors depend on population density; others do not.*

11. c. Growth of spotted owl populations appears to depend on the number and location of old-growth forests, the species' preferred habitat. For more information, see Section 22.5, Patterns of Population Growth.

12. True. If human birth rates dropped immediately to a level that would allow replacement but not increase, the population would continue to grow for at least 60 years because of the huge number of children who have not yet reproduced. For more information, see Section 22.5, Patterns of Population Growth.

13. e. All of these factors increase survivability in one way or another. The end result is more births and a lower death rate, both of which contribute to population increases. For more information, see Section 22.4, *Logistic population growth is the norm in the real world.*

14. True. Sustainable use of resources does not deplete the resources or seriously damage the environment. In the past, humans have used resources in a nonsustainable manner. For more information, see Section 22.4, *Logistic population growth is the norm in the real world.*

15. c. carrying capacity. For more information, see Section 22.4, *Logistic population growth is the norm in the real world.*
d. population size. For more information, see Section 22.1, What Is a Population?
a. exponential growth. For more information, see Section 22.3, Exponential Growth.
b. S-shaped growth curve. For more information, see Section 22.4, *Logistic population growth is the norm in the real world.*

e. population density. For more information, see Section 22.1, What Is a Population?

Conceptual Understanding

1. b. Both populations showed exponential growth, exceeded their carrying capacities, and then crashed. For more information, see Section 22.4, *Logistic population growth is the norm in the real world*, and Figure 22.8, Boom and Bust.

2. False. Natural populations are often difficult to define, and the definition may depend on the organism and particular research question. For more information, see Section 22.1, What Is a Population?

3. c. An increase from zero to a coverage of over 243,000 square kilometers in about 90 years easily qualifies as exponential growth. For more information, see Section 22.3, Exponential Growth.

4. a. Unlike natural systems, closed laboratory systems have no immigration or emigration. Space and nutrients usually are limited in both closed and natural systems. For more information, see Section 22.4, *Logistic population growth is the norm in the real world.*

5. b. If the factors that remove individuals from the population exceed those that add individuals, the population will decline. For more information, see Section 22.2, Changes in Population Size.

6. a. This is an environmental "fact of life" for all organisms in all habitats. For more information, see Section 22.1, What Is a Population?; Section 22.4, *Logistic population growth is the norm in the real world*; and Section 22.3, Exponential Growth.

7. True. Although it is impossible to predict accurately the maximum carrying capacity for a highly adaptable species such as humans, most experts agree that we are pushing the limits of Earth's ability to support our growing population. For more information, see Section 22.4, *Logistic population growth is the norm in the real world.*

8. False. A decrease in carrying capacity means that the habitat can support fewer individuals. For more information, see Section 22.4, *Logistic population growth is the norm in the real world.*

9. e. All of these practices lead to some form of environmental degradation, which is not a sustainable strategy for maintaining a healthy human population. For more information, see the chapter's Applying What We Learned article, "What Does the Future Hold?"

10. a. Americans are enormous consumers of energy and natural resources, far in excess of our percentage of representation in the world's human population. For

more information, see the Biology Matters box, "How Big Is Your Ecological Footprint?"

11. b. To be more precise, 8.3 billion. For more information, see the Applying What We Learned article, "What Does the Future Hold?"

12. b. Bald eagles are an example of a species that recovered from human interference. Several decades ago, our use of DDT in pesticides caused reproductive failure in eagles and other predatory birds. Banning DDT has led to the substantial recovery of their populations. For more information, see Section 22.5, Patterns of Population Growth, and Figure 22.11, Recovery of Bald Eagle Populations.

CHAPTER 23 | Ecological Communities

GETTING STARTED

Below are a few questions to consider before reading Chapter 23. These questions will help guide your exploration and assist you in identifying some of the key concepts presented in this chapter.

1. What term do ecologists use to describe an interaction that benefits both species?

2. How would extinction of the yucca plant affect the yucca moth?

3. What are the three groups of consumers that exploit other species for food?

4. Why are the spines on cacti that have been grazed longer than those on the plants that have not?

5. Why are many of the frogs found in the tropics so brightly colored?

6. What behavioral strategy do pigeons use to reduce goshawk predation?

7. Why was the American chestnut virtually defenseless against the fungus that causes chestnut blight?

8. In what circumstances will character displacement occur?

A GUIDE TO THE READING

The following concepts typically give students the most difficulty when exploring the content in Chapter 23 for the first time. For each concept, one or more references have been identified that may help you gain a better understanding of these potentially problematic areas.

Mutualism

Interactions among organisms affect individuals, populations, communities, and ecosystems. Ecologists classify these interactions by whether the interaction is beneficial or harmful to each of the interacting species. The key to understanding this classification is understanding that ecologists use the terms "beneficial" and "harmful" in the context of reproductive success. Any change that improves an organism's chances of mating or extends its life span would be considered beneficial. The death of an organism before it reproduces is, of course, the ultimate harm. Many varieties of mutualism occur. One of the most common examples is that of the bacteria that inhabit the digestive systems of animals and assist in the digestive process. As you read about other examples of mutualism, such as seed dispersal and pollination, evaluate each for the principles just discussed. Several misconceptions about mutualism are common. You might think, for example, that the relationship benefits both species equally, but this is generally not correct. It is common for each participant to experience some cost because of the relationship. Mutualism simply occurs when the benefits are greater than the costs. Some mutualistic relationships have become surprisingly specialized. In the most extreme examples, each participant is entirely dependent on the other, and neither can use an otherwise environmentally suitable habitat unless the other is present. As a result, mutualisms can influence the distribution and abundance of species. Similarly, should one species become locally or globally extinct, the other species will experience the same fate.

For more information on this concept, be sure to focus on

- Section 23.1, *There are many types of mutualisms*
- Section 23.1, *Mutualists are in it for themselves*
- Section 23.1, *Mutualisms can determine the distribution and abundance of species*

140 | *Chapter 23*

- Figure 23.3, Behavioral Mutualism
- Figure 23.4, Pollinator Mutualism
- Figure 23.5, The Home a Mutualism Built

Exploitation

Exploitation includes those interactions where one species benefits at the expense of another. Virtually all consumers, defined as organisms that are unable to produce their own food, are dependent on exploitation, making it one of the most common interactions in nature. Even those consumers that appear to be feeding on organic debris will be consuming the bacteria that first colonized the item. Natural selection has equipped the majority of prey species with a variety of defenses that work to reduce the harmful effects of exploitation. Induced defenses are those whose activity is stimulated by an attack. Plants often produce toxic chemicals only after grazing begins. Consider the elaborate response of the human immune system to infection by pathogenic bacteria or parasites, as discussed in Chapter 32. Prey defenses create an opportunity for changes within the consumer. For those species with long-term associations, many cycles of adaptation may have taken place, resulting in some unusual specializations.

Consider the rough-skinned newt. Its skin glands produce a toxin so potent that it is considered one of the most poisonous organisms on Earth. Only one predator has evolved the ability to overcome this toxin: the common garter snake. High toxicity is a personally marginal defense if the individual must die for it to be useful. The majority of toxic prey species advertise their toxicity through warning colorations. Bright, bold, or otherwise, highly visible patterns serve to communicate to potential consumers that the prey item will be distasteful or harmful. A variety of additional strategies that do not involve toxicity have evolved within prey populations. Consider the herding, flocking, or schooling behavior of some animals, birds, and small fish. Despite the fact that a group is more easily located than a single individual, the chances of any individual surviving an attack are much greater when within the group. The development of defensive and offensive strategies by natural selection is slow, often requiring dozens or hundreds of generations. The introduction of prey or consumer species into an ecosystem is often accompanied by extensive disruption. You read earlier about the introduction of the zebra mussel, a prey species growing without the control exerted by its natural predators. This chapter describes the effect of chestnut blight, a fungal consumer introduced to a prey population, the American chestnut, which lacked any natural defenses. As has been true in many other similar situations, the American chestnut is now virtually extinct where its distribution overlaps that of the fungus.

For more information on this concept, be sure to focus on

- Section 23.1, *In exploitation, one member benefits while another is harmed*
- Figure 23.10, Safety in Numbers

Competition

Competition is considered to be an interaction where both participants are harmed. There are two main types of competition: (1) In interference competition, one species is prevented from using a resource by the actions of the competitor species. The decline of coyotes following the reintroduction of wolves in Yellowstone National Park provides a good example. Although not eating the same foods as coyotes, wolves are highly territorial. Their territories left few spaces where coyotes could hunt without the risk of attack from wolves. Although any given wolf territory contains an abundance of unused food items suitable for the coyote, the action of the wolves interferes with the coyote feeding, and the latter population has declined. (2) In exploitation competition, the resource is used by both species, and its use by one competitor prevents its use by the other. Competition for space is one of the most common examples of exploitation competition. On rocky intertidal shorelines throughout the world, competition occurs for attachment sites. Ecologists had long wondered why so many intertidal organisms lived in horizontal bands with clear boundaries between them. Research has shown that the lower limit of the distribution of one species is almost always the result of a biological interaction with the species below, typically the better competitor.

For more information on this concept, be sure to focus on

- Section 23.1, *In competition, both species are negatively affected*
- Figure 23.12, What Keeps Them Apart?

Communities Change over Time

All communities change over time. One source of change involves biological interactions and is called "succession." In succession, a series of communities occupies a site in a somewhat orderly and predictable fashion. In Chapter 22 we learned that living organisms can alter the physical environment. An existing plant community exerts powerful effects on the structure of the soil. Its roots penetrate the soil, its leaves and roots may decompose and add organic material, and the shade it creates may reduce evaporation and raise moisture levels. With time, these improvements may help competitors displace some species, and the composition of the community will change. Ecologists distinguish between primary succession, which usually begins on a newly created habitat, and secondary succession, which occurs after a disturbance removes the existing community but leaves intact the soil and the roots and seeds it contains. Change in communities may be unrelated to biological processes. We also learned in Chapter 22 that climate is one of the most influential factors on community and biome development. Earth's climate regularly cycles through long-scale periods of cooling and warming. We are now in a relatively warm interglacial period. Should the climate run through its cycle as it has for the past 400,000 years, we can anticipate another cooling trend and ice age in the near future. Even if climatic cycles were to cease, habitats would continue to experience slow climatic change because of continental drift (Chapter 20). The slow movement of the continental land masses eventually moves them to different latitudes and the climates associated with them. A slow northward drift has moved North America during the past 135 million years from a semitropical location to its current temperate and arctic position.

For further information on this concept, be sure to focus on

- Section 23.3, *Succession establishes new communities and replaces disturbed communities*
- Figure 23.16, Succession
- Section 23.3, *Communities change as climate changes*
- Figure 23.18, Climate Change, Communities Change

TYING IT ALL TOGETHER

Several concepts presented in this chapter build on those presented in previous chapters and may also be revisited and discussed in greater detail in subsequent chapters, including

Mutualism

- Chapter 3—Section 3.4, Fungi: A World of Decomposers
- Chapter 35—Section 35.4, *Roots absorb nutrients from the soil through cells and along cell walls*

Exploitation

- Chapter 17—Section 17.2, Mechanisms of Evolution
- Chapter 24—Section 24.2, Energy Capture in Ecosystems
- Chapter 32—Section 32.4, Third Line of Defense: The Adaptive Immune System

Competition

- Chapter 22—Section 22.4, Logistic Growth and the Limits on Population Size

PRACTICE QUESTIONS

Factual Knowledge

1. Pollinator mutualisms are special interactions involving _____, which receive food or a place to lay eggs, and _____, which receive pollen from others of their kind.
   a. insects; plants
   b. plants; insects
   c. parasites; plants
   d. predators; plants
   e. none of the above

2. Mutualists are altruistic in that they react to the other species in ways that do *not* benefit themselves. (True or False)

3. In a mutualistic relationship, only the species directly involved benefit. (True or False)

4. Warning coloration is a result of
   a. consumers placing selective pressure on a species.
   b. victims evolving faster than consumers.
   c. behavior changing the physiology of an organism.
   d. all of the above
   e. none of the above

5. Consumers can alter which of the following characteristics of victims?
   a. distribution
   b. abundance
   c. behavior
   d. all of the above
   e. none of the above

6. Group interaction is often an alteration of behavior to
   a. reduce selective pressures.
   b. avoid consumption by another species.
   c. speed character displacement.
   d. all of the above
   e. none of the above

7. Species compete indirectly for shared resources in _____ competition.
   a. interference
   b. exploitative
   c. interrelation
   d. infringement
   e. none of the above

8. The level of competition between species depends on
   a. availability of resources.
   b. population density.
   c. group interaction of organisms.
   d. all of the above
   e. none of the above

9. Two species of Galápagos finches have very similar beak sizes on islands where only one species is found, but different beak sizes on islands where they are found together. This is an example of
   a. convergent evolution.
   b. character displacement.
   c. competitive fitness.
   d. ephemeral differences.
   e. all of the above

10. Food chains differ from food webs in that
    a. food chains are a single sequence of who eats whom in a community.
    b. food webs represent the complex interactions among food chains.
    c. food chains better represent the entire community.
    d. both a and b
    e. both a and c

11. Keystone species are always predators. (True or False)

12. Overgrazing, overlogging, overburning, and overfarming are all examples of human impact on
    a. communities.
    b. succession.
    c. consumers.
    d. all of the above
    e. none of the above

13. "Secondary succession" refers to the second phase of succession that occurs after a disaster. (True or False)

14. Which of the following is a symbiotic relationship?
    a. gut inhabitant mutualism
    b. parasitism
    c. intraspecific competition
    d. both a and b
    e. none of the above

15. Match each term with the best description.
    __ behavioral mutualism
    __ character displacement
    __ interference competition
    __ parasite
    __ predator
    a. forms of competing species evolve differently over time to lessen competition
    b. the member of a consumer-victim relationship that hunts and kills prey
    c. a species altering its behavior to benefit another in order to gain benefits for itself
    d. competition through preventing access to resources
    e. the member of a consumer-victim relationship that attains its resources by living off of a host

Conceptual Understanding

1. Two organisms share the same habitat and consume from the same food source. Neither prevents the other from obtaining the food source. This is an example of _____ competition.
   a. interference
   b. exploitative
   c. intimate
   d. all of the above
   e. none of above

2. In lichens, fungi and cyanobacteria live together without causing significant harm to one another. Attempts to grow either the cyanobacteria or the fungi independently result in the death of the organism. This interaction is an example of
   a. parasitism.
   b. mutualism.
   c. competition.
   d. all of the above
   e. none of the above

3. Gazelles in a group have a 75 percent chance of escaping hungry lions. Solitary gazelles have only an 8 percent chance of escaping. In this case, _____ behavior is advantageous for avoiding predation.
   a. mutualistic
   b. individual
   c. group
   d. all of the above
   e. none of the above

4. Barnacles on the Scottish shore, as shown in Figure 23.12 of your textbook, illustrate _____ competition.
   a. interference
   b. exploitative
   c. intimate
   d. all of the above
   e. none of the above

5. Lizards on a chain of islands prefer to eat the berries of the Make-Believe bush. Two species of lizards compete for this resource. It is noticed that on Fantasy Island the two species of lizards differ in the length of their legs, with one species grazing on berries at the bottom of the bush and the other species grazing on the top of the bush. However, on the other two islands of the chain, where only one or the other species is present, they appear to have legs of the same length. This is an example of
   a. character development.
   b. character displacement.
   c. character morphogenesis.

d. all of the above
e. none of the above

6. During the last century, a volcanic eruption destroyed an entire island. Several years later a new island, sterile and devoid of life, was born on the spot where the original island had disappeared. Scientists were able to view the island before colonization began and as it established itself. These scientists were witnessing
   a. primary succession.
   b. secondary succession.
   c. foundation building.
   d. keystone placement.
   e. none of the above

7. Changes in the makeup of communities in Australia result from
   a. succession.
   b. climate change.
   c. human impact.
   d. all of the above
   e. none of the above

8. If a new predator is introduced to an area with a type of prey that is limited in number, the prey species is likely to
   a. decrease, and possibly become extinct.
   b. increase.
   c. remain stable.
   d. initially increase, then decrease.
   e. none of the above

9. In a pond, you notice a snapping turtle, tadpoles, algae, various species of aquatic insects, and many fish. This pond community represents a food
   a. chain.
   b. web.
   c. pyramid.
   d. all of the above
   e. none of the above

10. You have a pet snake that is fed rats. The rats eat a grain-based diet. The snake is a
   a. primary consumer.
   b. producer.
   c. secondary consumer.
   d. herbivore.
   e. none of the above

11. Mutualistic relationships
   a. do not have costs.
   b. evolve when the benefits of the interaction outweigh the costs.
   c. are rare.
   d. are limited to plants and animals.
   e. none of the above

12. Which of the following statements about parasites is true?
   a. They often harm but don't immediately kill their hosts.
   b. They can alter the behavior of their hosts.
   c. They may kill their host if this promotes transmission to another host.
   d. all of the above
   e. none of the above

RELATED ACTIVITIES

- Make a list of 10 common species interactions in a habitat near your school. Next to each one, identify which type of interaction it is and the role each member plays in it.
- For each type of interaction discussed in this chapter, think of an example that involves humans. Briefly describe the interaction and the other organisms involved.
- Using your library or the Internet, find an example of a situation in which competition or consumption has led to the extinction of a species. Write a one-page essay describing the example.
- Using the library or the Internet, investigate how prey species can benefit from living in groups. Then examine whether living in groups makes some animals more susceptible to predation. Outline your findings in a short essay.
- Develop a brochure that highlights what your community is doing to minimize the impact humans are having on the environment.
- Investigate your state's land preservation policies. What measures are being taken to set aside lands for preservation, and what has been the impact of these measures on the communities of organisms residing in the protected areas?

ANSWERS AND EXPLANATIONS

Factual Knowledge

1. a. In pollinator mutualisms, insects pollinate plants in exchange for food or other rewards. For more information, see Section 23.1, *There are many types of mutualism.*
2. False. Mutualists receive benefits for themselves. Without these benefits, the interaction would not exist. For more information, see Section 23.1, *Mutualists are in it for themselves.*
3. False. In addition to the species directly involved in a mutualistic relationship, other organisms may also benefit. For example, coral reefs result from the

mutualistic relationship between corals and photosynthetic algae, and these reefs provide a home to many other organisms. For more information, see Section 23.1, *Mutualism can determine the distribution and abundance of species.*

4. a. Those with warning coloration are less likely to get eaten, and their genes are passed on. For more information, see Section 23.1, *Consumers and their food organisms can exert strong selection pressure on each other.*

5. d. Consumers can alter the distribution, abundance, and behavior of their victims. For more information, see Section 23.1, *Consumers can restrict the distribution and abundance of their food organisms* and *Consumers can alter the behavior of exploited organisms.*

6. b. Groups have better detection and warning mechanisms than individuals do. For more information, see Section 23.1, *Consumers can alter the behavior of exploited organisms.*

7. b. The resource is available to both competitors. For more information, see Section 23.1, *In competition, both species are negatively affected.*

8. d. Limitation of resources and higher population density will result in greater competition among organisms. Group interactions may restrict or increase competition, depending on the circumstances. For more information, see Section 23.1, *In competition, both species are negatively affected.*

9. b. Character displacement is the altering of a character in areas of high competition so that other resources may be exploited. For more information, see Section 23.1, *Competition can increase the differences between species.*

10. d. The complex relationships among food chains are best described using a food web. For more information, see Section 23.2, *Food chains transfer nutrients through the community.*

11. False. Although some keystone species are predators, others are producers, herbivores, or pathogens. For more information, see Section 23.2, *Keystone species have profound effects on communities.*

12. a. Humans are responsible for some, if not all, of the largest impacts on communities. For more information, see Section 23.4, *People can cause long-term damage to communities.*

13. False. Secondary succession refers to the recolonization of an area after a disturbance. For more information, see Section 23.3, *Succession establishes new communities and replaces disturbed communities.*

14. d. Gut inhabitant mutualism and parasitism are symbiotic relationships because in each relationship two different species live in direct contact. For more information, see Section 23.1, *There are many types of mutualism.*

15. c. behavioral mutualism. For more information, see Section 23.1, *There are many types of mutualism.*
    a. character displacement. For more information, see Section 23.1, *Competition can increase the differences between species.*
    d. interference competition. For more information, see Section 23.1, *In competition, both species are negatively affected.*
    e. parasite. For more information, see Section 23.1, *In exploitation, one member benefits while another is harmed.*
    b. predator. For more information, see Section 23.1, *In exploitation, one member benefits while another is harmed.*

Conceptual Understanding

1. b. In exploitative competition, competing species all have access to the resource. For more information, see Section 23.1, *In competition, both species are negatively affected.*

2. b. Lichens exemplify mutualism. For more information, see Section 23.1, *There are many types of mutualism.*

3. c. Predation is a driving force behind group behavior in this case. For more information, see Section 23.1, *Consumers can alter the behavior of exploited organisms.*

4. a. One group of barnacles interferes with the other's access to resources. For more information, see Section 23.1, *Competition can limit the distribution and abundance of species.*

5. b. These similar species show divergence of characters as a result of competition. For more information, see Section 23.1, *Competition can increase the differences between species.*

6. a. A barren, uncolonized island is a textbook example of primary succession. For more information, see Section 23.3, How Communities Change over Time.

7. d. All of the factors listed contribute to community change. For more information, see Section 23.3, *Communities change as climates change.*

8. a. The initial result will be a reduction in the number of prey species. The prey may well become extinct, as happened to many species of birds in New Zealand following the introduction of house cats. For more information, see Section 23.1, *Consumers and their food organisms can exert strong selection pressure on each other.*

9. b. This is definitely a food web with many food chains represented. For more information, see Section 23.2, *Food chains transfer nutrients through the community*, and Figure 23.14, A Marine Food Web.

10. c. In this example, the rat is a primary consumer that

eats the producer, and the snake is a secondary consumer because it eats the rat. For more information, see Section 23.2, *Food chains transfer nutrients through the community.*

11. b. Mutualistic relationships are usually not cost free; instead, such relationships evolve when the benefits exceed the costs for the species involved. In addition, mutualisms are very common and are not limited to plants and animals. For more information, see Section 23.1, *There are many types of mutualism* and *Mutualists are in it for themselves.*

12. d. Parasites harm their hosts and may alter their behavior. Some parasites kill their host if killing promotes transmission to the next host in their life cycle. For more information, see Section 23.1, *Consumers and their food organisms can exert strong selection pressure on each other* and *Consumers can alter the behavior of exploited organisms.*

CHAPTER 24 | Ecosystems

GETTING STARTED

Below are a few questions to consider before reading Chapter 24. These questions will help guide your exploration and assist you in identifying some of the key concepts presented in this chapter.

1. How did New York City decide to correct its problem with poor water quality?

2. Why are hawks and owls so much less common than the rodents they eat?

3. How is agriculture in the Midwest related to the dead zone in the Gulf of Mexico?

4. If energy can neither be made nor destroyed, why can't it be recycled in an ecosystem?

5. What two factors best predict the level of net primary productivity (NPP) for any region?

6. In what way does the cycling of phosphorus differ from the cycling of all the other major nutrients?

7. What pollutant is responsible for most acid rain?

8. How successful were the scientists that built Biosphere II in designing an artificial ecosystem?

A GUIDE TO THE READING

The following concepts typically give students the most difficulty when exploring the content in Chapter 24 for the first time. For each concept, one or more references have been identified that may help you gain a better understanding of these potentially problematic areas.

Ecosystems

An ecosystem consists of a community of organisms along with the physical environment in which those organisms live. Two primary activities characterize an ecosystem: the unidirectional flow of energy and the recycling of nutrients. Ecosystems often have no physical boundaries and vary in size from areas as small as mud puddles to entire oceans. With no specific size requirement and no discernable boundaries, it might seem impossible to identify an ecosystem. Rather than using physical features, ecologists delineate ecosystems using functional criteria. All the organisms that utilize the energy and nutrients from a single set of producers belong to the ecosystem. Organisms obtaining their energy and nutrients elsewhere belong to a different ecosystem. Of course, such a simple explanation is rarely suited to the complexities of nature. Many organisms are highly mobile (the Arctic tern, for example, flies 22,000 miles annually) and frequently move between ecosystems.

For more information on this concept, be sure to focus on

- Section 24.1, How Ecosystems Function: An Overview
- Figure 24.2, How Ecosystems Work

Energy Capture in Ecosystems

All living organisms require energy. Consumers obtain energy by digesting the complex chemical compounds they obtain from their food. Producers, however, collect their energy from nonliving sources, primarily the sun. Producers use solar energy, carbon dioxide, and water (Chapter 8) to photosynthesize the energy-rich compounds that will eventually nourish their bodies, primary consumers directly, and all other consumers indirectly. Because photosynthesis is the essential first step in ecosystem function, ecologists find it useful to characterize different ecosystems by the amount of photosynthesis that takes place within them. Total photosynthesis, however, would be a misleading measurement, because a portion of the captured energy is used by the producer. Subtracting this quantity from total photosynthesis gives a value called "net primary productivity," or NPP. It is the standard measure of ecosystem productivity and has been used worldwide to compare one ecosystem with another. Patterns of NPP are predictably variable. For example, because the intensity of sunlight varies with the latitudes, you might predict high NPP near the equator, with declines as latitude increases. Of course, photosynthesis also requires water.

In Chapter 33 we will learn that Earth's atmospheric convection cells produce complex patterns of rainfall. Areas receiving ample sunlight and abundant rain, like the equator, do experience high NPP. Areas with abundant sunlight and little rain have a low NPP. The NPP in aquatic biomes is less predictable. Aquatic organisms are able to use much lower light intensities than terrestrial plants because light penetrates the water column poorly. Productivity in aquatic biomes is often more dependent on the availability of dissolved nutrients. Rivers and streams typically deliver these, creating highly productive wetlands such as estuaries and marshes where the nutrients begin to accumulate. Human activity can significantly increase or decrease NPP. Nutrient enrichment from agricultural runoff or sewage outflows can produce almost unimaginable increases in productivity in aquatic ecosystems. Growth can be so abundant that the water becomes opaque from the organisms it contains. Recall that exponential growth seldom lasts for long. These organisms die rapidly, followed by bacterial decomposition that can entirely deplete the dissolved oxygen within the area. Oxygen depletion has now become an apparently permanent feature of the Gulf of Mexico. Such massive disruptions to ecosystem function are almost always detrimental.

For more information on this concept, be sure to focus on

- Section 24.2, Energy Capture in Ecosystems
- Figure 24.13, The Dead Zone

Nutrient Cycles

Ecologists use the term "nutrient" differently than most people. To an ecologist, nutrients are the various atoms and molecules needed to build the body of an organism. Nutrients may or may not provide energy. Thus, the most important nutrients tend to be simple chemicals like carbon dioxide (CO_2), nitrogen dioxide (NO_2-), and potassium ion ($K+$). Producers obtain their nutrients from the soil, water, or air. Consumers must eat either producers or other consumers. The transfer of a nutrient between organisms and the physical environment is termed a "nutrient cycle." Nutrients may be difficult for producers to obtain, and shortages often limit the

productivity of an ecosystem. The time required for a nutrient to cycle between a producer, the physical environment, and another producer depends on whether the nutrient cycles in an atmospheric cycle or a sedimentary cycle. In atmospheric cycling, the nutrient is a gas. It moves freely as the atmosphere circulates and tends to be available to almost all ecosystems. Nutrients that cycle in the atmosphere are rarely limiting. In the sedimentary cycle, soluble nutrients are transported by river to the sea, where they eventually become incorporated into marine sediments and become sedimentary rock. Over millions of years, these sediments may be uplifted, exposed to weathering, and eroded, so that the nutrients are allowed to reenter the biosphere. Because this time interval is so long, these nutrients have the potential to become limiting factors. Phosphorus is the only major nutrient with a sedimentary cycle. Human activity commonly disrupts nutrient cycling, leading to surplus nutrients in some ecosystems and depletions in others. One of the earliest studies of nutrient cycling demonstrated that forest clear-cutting significantly accelerates the rate of nitrogen loss from forest soils. The practice of wastewater treatment often adds phosphorus to lakes and rivers, contributing to eutrophication and a deterioration of water quality.

For more information on this concept, be sure to focus on

- Section 24.4, Biogeochemical Cycles
- Figure 24.8, Nutrient Cycling
- Figure 24.9, The Carbon Cycle
- Figure 24.10, The Nitrogen Cycle
- Figure 24.11, The Sulfur Cycle
- Figure 24.12, The Phosphorus Cycle

TYING IT ALL TOGETHER

Several concepts presented in this chapter build on those presented in previous chapters and may also be revisited and discussed in greater detail in the subsequent chapters, including

Ecosystems

- Chapter 23—Section 23.1, *In exploitation, one member benefits while another is harmed*

Energy Capture in Ecosystems

- Chapter 9—Section 9.3, Photosynthesis: Capturing Energy from Sunlight

Nutrient Cycles

- Chapter 22—Section 22.4, *Growth is limited by essential resources and other environmental factors*

- Chapter 25—Section 25.3, Changes in Global Nutrient Cycles
- Chapter 35—Section 35.4, *Plants need mineral nutrients to grow*

PRACTICE QUESTIONS

Factual Knowledge

1. To continue to provide high-quality water for its residents, New York City decided to
   a. build desalinization plants along its Atlantic shoreline.
   b. construct water treatment facilities for its entire water supply.
   c. purchase the land bordering its watersheds to prevent development and pollution from further reducing water quality.
   d. construct rainwater capture basins to collect and store rainwater.
   e. all of the above

2. The greatest amount of energy is available at the _____ of a food chain.
   a. bottom
   b. top
   c. middle
   d. top and middle
   e. none of the above

3. The amount of nutrients available on Earth is
   a. increasing.
   b. fixed.
   c. decreasing.
   d. variable.
   e. none of the above

4. All life on Earth depends on _____ energy.
   a. solar
   b. heat
   c. chemical
   d. synthetic
   e. none of the above

5. Because of limitations on resources, organisms must _____ materials in order to survive.
   a. stockpile
   b. reinvent
   c. reintroduce
   d. recycle
   e. none of the above

6. Ecosystems are _____ because resources _____ move from one ecosystem to another.
   a. open; can
   b. closed; cannot
   c. one-way; can
   d. one-way; cannot
   e. none of the above

7. Ecosystem ecologists measure the
   a. diversity of species in an ecosystem.
   b. movement of energy and materials into and out of an ecosystem.
   c. natural resources available to a system.
   d. all of the above
   e. none of the above

8. Energy is recycled. (True or False)

9. Net primary productivity (NPP) of a system refers to the amount of energy
   a. obtained through photosynthesis.
   b. lost through cellular respiration.
   c. obtained through photosynthesis minus the amount lost through cellular respiration.
   d. stored as chemical energy.
   e. none of the above

10. NPP depends only on the amount of sunlight available. (True or False)

11. The majority of available nutrients in terms of biomass go to
    a. producers.
    b. consumers.
    c. decomposers.
    d. producers and consumers combined.
    e. none of the above

12. The primary difference between sedimentary and atmospheric cycles is that in _____ cycles the nutrient does not _____.
    a. sedimentary; leave the terrestrial environment
    b. sedimentary; leave the aquatic environment
    c. atmospheric; leave the aquatic environment
    d. sedimentary; enter the atmosphere
    e. none of the above

13. Acid rain results when humans put excess amounts of _____ into the atmosphere.
    a. phosphorus
    b. sulfur
    c. potassium
    d. hydrogen
    e. lithium

14. Match each term with the best description.
    __ trophic level
    __ ecosystem
    __ secondary productivity
    __ biomass
    __ nutrient cycle

a. unit of measure of NPP
b. another name for steps in a food chain
c. all of the organisms and the environments in which they live
d. rate of new biomass production by consumers
e. cyclical movement of a nutrient between organisms and the physical environment

Conceptual Understanding

1. Secondary productivity depends on primary net productivity. (True or False)

2. Energy pyramids are used to represent energy transfer in an ecosystem because energy is _____ between each trophic level.
   a. gained
   b. lost
   c. conserved
   d. either conserved or gained
   e. neither lost nor gained

3. Rapid cycling of nutrients occurs in
   a. sedimentary nutrient cycles.
   b. atmospheric nutrient cycles.
   c. NPP.
   d. all of the above
   e. none of the above

4. Humans benefit from ecosystems because ecosystems provide
   a. buffers from natural disasters such as floods.
   b. maintenance of a clean water supply.
   c. climate moderation.
   d. all of the above
   e. none of the above

5. A pasture of 40 hectares has a biomass of 40 kilograms per hectare, or roughly 1 million calories of energy. Cattle grazing on this pasture are sold for beef. About how many calories of energy from the pasture reach human consumers of the beef?
   a. 10,000
   b. 1,000
   c. 100,000
   d. 1 million
   e. 100

6. Phosphorus is a sedimentary nutrient. This means it
   a. cycles very slowly.
   b. never enters the atmosphere.
   c. settles as sediment in the ocean.
   d. all of the above
   e. none of the above

7. Decomposers such as fungi and bacteria are vital to the ecosystem because they

a. are the primary organic recyclers.
b. are the terminal energy consumers.
c. keep populations in check.
d. both a and b
e. both a and c

8. Energy flow through an ecosystem differs from nutrient flow in that
   a. energy flow is linear.
   b. nutrient flow is linear.
   c. energy flow is cyclical.
   d. energy builds as it flows through the ecosystem.
   e. none of the above

9. The majority of the energy available in an ecosystem is actually lost as heat energy. (True or False)

10. Conifers are especially sensitive to acid rain because their needles literally rake moisture out of clouds. Because of acid rain, Mount Mitchell, the highest peak in the eastern Appalachians, has seen dramatic loss of pine forest. The damage on this mountain most likely stems from
    a. droughtlike conditions in the region.
    b. excessive sulfur in the atmosphere due to industrialization along the eastern seaboard.
    c. loss of biomass throughout the Appalachians.
    d. all of the above
    e. none of the above

11. Estuaries have one of the highest levels of NPP because
    a. they are relatively shallow, with good light penetration.
    b. nutrients transported by rivers enter the estuary before reaching the ocean.
    c. the action of the tides slows the deposition of nutrients into the bottom sediments.
    d. estuaries are often warmer than the adjacent river or nearby ocean.
    e. all of the above

12. Food chains of more than five trophic levels are quite common. (True or False)

RELATED ACTIVITIES

- Investigate what is being done globally and by individual nations to address issues relating to acid rain. How is the United States attempting to curb acid rain? Write a one-page essay describing your findings.
- Identify two plots of at least one square meter on your campus with different physical features like soil or exposure to sunlight. With the help of your instructor, devise a method to compare NPP for the sites. Write a

brief summary of your findings, proposing a testable hypothesis that could account for any differences between the sites.

• Read or reread Rachel Carson's *Silent Spring*. Write a one-page essay that identifies which of her predictions have come true and that discusses the progress we have or have not made in the areas she addressed.

ANSWERS AND EXPLANATIONS

Factual Knowledge

1. c. The high cost of alternatives helped New York City realize that protecting the natural system that provided clean water was cost effective. For more information, see the chapter's Biology Matters box, "Is There a Free Lunch? Ecosystems at Your Service."

2. a. Producers capture solar energy. At each subsequent trophic level, only about 10 percent of the energy is transferred. For more information, see Section 24.3, *An energy pyramid shows the amount of energy transferred up a food chain*.

3. b. The fixed nature of nutrients means that they must be recycled. For more information, see Section 24.1, How Ecosystems Function: An Overview.

4. a. Producers convert solar energy to usable energy by way of the remaining trophic levels. For more information, see Section 24.2, Energy Capture in Ecosystems.

5. d. Nutrient cycling is necessary for life on Earth. For more information, see Section 24.1, How Ecosystems Function: An Overview.

6. a. Energy and materials can pass from one ecosystem to another. For more information, see Figure 24.2, How Ecosystems Work, and Section 24.1, How Ecosystems Function: An Overview.

7. b. By means of uniform measurements like biomass, it is possible to make comparisons between ecosystems. For more information, see Section 24.1, How Ecosystems Function: An Overview.

8. False. Energy moves through an ecosystem. For more information, see Section 24.3, *An energy pyramid shows the amount of energy transferred up a food chain*.

9. c. The "net" of anything is equal to that which was acquired minus the cost of operation to acquire it. For more information, see Section 24.2, *The rate of energy capture varies across the globe*.

10. False. Producers can't produce without adequate water and CO_2 and favorable environmental conditions. For more information, see Section 24.2, *The rate of energy capture varies across the globe*.

11. c. Decomposers take in more than 50 percent of available biomass. For more information, see Section 24.3, *Secondary productivity is highest in areas of high NPP*, and Figure 24.7, Give Thanks to the Decomposers.

12. d. This characteristic makes sedimentary nutrient cycling necessarily slow. For more information, see Section 24.4, Biogeochemical Cycles.

13. b. Excess sulfur is common in industrialized nations and causes acid rain. For more information, see Section 24.4, *Sulfur is one of several important nutrients with an atmospheric cycle*, and Figure 24.11, The Sulfur Cycle.

14. b. trophic level. For more information, see Section 24.3, *An energy pyramid shows the amount of energy transferred up a food chain*.

    c. ecosystem. For more information, see Section 24.1, How Ecosystems Function: An Overview.

    d. secondary productivity. For more information, see Section 24.3, *Secondary productivity is highest in areas of high NPP*.

    a. biomass. For more information, see Section 24.2, *The rate of energy capture varies across the globe*.

    e. nutrient cycle. For more information, see Section 24.4, Biogeochemical Cycles.

Conceptual Understanding

1. True. Consumers can produce only if they have producers to eat. For more information, see Section 24.3, *Secondary productivity is highest in areas of high NPP*.

2. b. Only 10 percent of the energy is transferred between trophic levels. For more information, see Section 24.3, *An energy pyramid shows the amount of energy transferred up a food chain*.

3. b. Atmospheric availability allows for more rapid nutrient cycling. For more information, see Section 24.5, *Human activities can alter nutrient cycles*.

4. d. With such benefits available, we should be responsible about our use of ecosystems. For more information, see the Biology Matters box, "Is There a Free Lunch? Ecosystems at Your Service," and Section 24.1, How Ecosystems Function: An Overview.

5. c. Only about 10 percent of the energy is transferred between trophic levels; thus, 100,000 calories are in the cows and available to human consumers. For more information, see Section 24.3, *An energy pyramid shows the amount of energy transferred up a food chain*.

6. d. Sedimentary nutrients cycle slowly because they settle on the bottom of the ocean and do not enter the

atmosphere. The upwelling of the ocean floor is needed for them to be released and made available for biological use. For more information, see Section 24.4, Biogeochemical Cycles.

7. a. Decomposers are vital to the health of the ecosystem because they prevent nutrients from being tied up in one place for extended periods. Energy is not recycled; vast quantities have become fossilized without any appreciable impact on ecosystem function. For more information, see Figure 24.2, How Ecosystems Work; Section 24.2, *Ecosystems depend upon energy captured by producers*; and Section 24.4, Biogeochemical Cycles.

8. a. Energy cannot be recycled. For more information, see Section 24.3, *An energy pyramid shows the amount of energy transferred up a food chain.*

9. True. A large portion of available energy is lost as heat. For more information, see Section 24.1, How Ecosystems Function: An Overview.

10. b. Excess sulfur is a primary cause of acid rain. For more information, see Section 24.4, *Sulfur is one of several important nutrients with an atmospheric cycle.*

11. e. Estuaries are somewhat protected, often warmed, and enriched with nutrients from the nearby terrestrial ecosystems, conditions that support high NPP. For more information, see Section 24.2, *The rate of energy capture varies across the globe.*

12. False. Because only 10 percent of energy is transferred from one trophic level to the next, most food chains have only a few levels. For more information, see Section 24.3, *An energy pyramid shows the amount of energy transferred up a food chain.*

CHAPTER 25 | Global Change

GETTING STARTED

Below are a few questions to consider before reading Chapter 25. These questions will help guide your exploration and assist you in identifying some of the key concepts presented in this chapter.

1. Why has the harvest of large predatory fish—like swordfish, shellfish, and white abalone—been in decline for the past 40 years?

2. When land is transformed, what does it become?

3. Why is the concentration of pollutants so much higher in the bodies of higher-order consumers than in the environments where they live?

4. What are synthetic chemicals, and why are organisms so poorly able to deal with them?

5. If nitrogen is a limiting resource, why have increased levels of nitrogen caused environmental problems?

6. What were the results of the Montreal Protocol?

7. In addition to carbon dioxide, what other greenhouse gases are serious environmental concerns?

8. Why is the conclusion that global temperatures are rising so controversial?

A GUIDE TO THE READING

The following concepts typically give students the most difficulty when exploring the content in Chapter 25 for the first time. For each concept, one or more references have been identified that may help you gain a better understanding of these potentially problematic areas.

A Worldwide Change in the Environment

Human activity has changed the biosphere substantially in the past 1,000 years. Collectively, these changes are referred to as "land transformations" when the land surface has changed and "water transformations" when aquatic biomes are affected. Abundant evidence documents these changes, and the total area affected is surprisingly large. Ecologists estimate that humans have substantially altered a third to a half of Earth's land surface. Water transformation has been more difficult to document, but declines in the worldwide harvest of fish and shellfish are unprecedented in modern times. Ecosystems within these transformed areas have been dramatically affected. In the United States, wetland acreage, for example, has been in a continuous decline in every state for the past 200 years. When not directly lost, many ecosystems have been extensively modified to produce the goods and services needed by a growing human population. Humans are currently estimated to control 30–35 percent of the world's net primary productivity (NPP). Diverting such a significant portion of an ecosystem's NPP greatly reduces the amount of land and resources available to other species.

For more information on this concept, be sure to focus on

• Section 25.1, Land and Water Transformation

Global Changes in Nutrients

The benefit of supplementing soil nutrients has long been known. In modern agriculture, the application of both synthetic chemicals, such as pesticides, and natural chemicals, such as phosphorus and nitrogen, has reached alarming proportions. Although it constitutes almost 80 percent of the atmosphere, atmospheric nitrogen cannot be used directly by producers. The conversion to a usable form, called "nitrogen fixation," is performed by certain species of bacteria. Currently, the amount of nitrogen fixation accomplished by human activities exceeds that of all natural sources combined. The consequences of changing the nitrogen cycle are far reaching. When nitrogen is a limiting factor, NPP increases dramatically. Not all members of the ecosystem benefit, however. Those able to use the extra nitrogen may outcompete other species. One of the most common outcomes of nutrient enrichment is an eventual reduction in the number of species within the ecosystem. This is of concern because ecosystems are most stable when the number of species is greatest. An additional concern has been intensifying after the initial identification of nitrogen enrichment in the 1960s. Synthetic chemicals and toxic metals have been introduced into food webs by human activities. However, organisms typically lack the capacity to rid their bodies of these substances, which are stored internally, and when consumed as prey they pass this pollution on to the predator in a process called "biomagnification." Higher-order consumers may have concentrations of pollutants that exceed environmental levels by a million–fold. Some long-lived carnivorous marine fish are so contaminated that eating them is a health risk.

For more information on this concept, be sure to focus on

- Section 25.2, Changes in the Chemistry of Earth
- Section 25.3, Changes in Global Nutrient Cycles
- Figure 25.4, Human Impact on the Global Nitrogen Cycle

Global Warming

Atmospheric scientists around the world are uniformly concerned about the increases in atmospheric carbon dioxide. Beginning in the 1700s, Earth's atmospheric levels of carbon dioxide have increased to those that have not occurred naturally for the past 420,000 years. Carbon dioxide is transparent to the solar energy that reaches the earth, but reacts strongly with the heat energy released from Earth's surface. Acting just like the glass windows in a car on a cold clear day, the heat energy is trapped, warming either the car's interior or Earth itself. Average global temperatures have been increasing slowly for as long as reliable records are available. In Chapter 21 we learned that temperature and rainfall provide the primary influences that dictate biome development. As global temperatures change, so will the patterns of atmospheric circulation and rainfall. Climatologists wonder if the new patterns of rainfall will continue to support agriculture in the areas where it now occurs. The soils that make agriculture possible have developed with thousands of years of climatic influence. There is a significant possibility that rainfall patterns might shift to areas with unsuitable soils. The potential for human impact alone is tremendous.

For more information on this concept, be sure to focus on

- Section 25.3, Changes in Global Nutrient Cycles
- Figure 25.6, Atmospheric CO_2 Levels Are Rising Rapidly
- Section 25.4, Climate Change

TYING IT ALL TOGETHER

Several concepts presented in this chapter build on those presented in previous chapters and may also be revisited and discussed in subsequent chapters, including

A Worldwide Change in the Environment

- Chapter 22—Section 22.4, Logistic Growth and the Limits on Population Size
- Chapter 23—Section 23.3, *Succession establishes new communities and replaces disturbed communities*
- Chapter 24—Section 24.2, *The rate of energy capture varies across the globe*

Global Changes in Nutrients

- Chapter 24—Section 24.5, *Human activities can alter nutrient cycles*

Global Warming

- Chapter 21— Section 21.2, Climate's Large Effect on the Biosphere

PRACTICE QUESTIONS

Factual Knowledge

1. According to ecologists' estimations, the amount of land surface that has been altered by humans is
   a. about one-fourth.
   b. between one-third and one-half.
   c. less than one-eighth.
   d. about three-fourths.
   e. undetermined at this time.

2. Humans control approximately _____ percent of the world's NPP.
   a. 10–15
   b. 20–25
   c. 30–35
   d. 40–45
   e. 50–55

3. Humans change how chemicals are cycled through ecosystems by
   a. adding synthetic chemicals to the atmosphere.
   b. releasing excessive amounts of natural chemicals through industrial processes.
   c. using chemicals to increase farm productivity.
   d. all of the above
   e. none of the above

4. The majority of available nitrogen on Earth is in the form of nitrogen gas, which cannot be used by living organisms. For it to be usable, nitrogen must be converted to
   a. nitrate.
   b. nitrite.
   c. ammonium.
   d. both a and b
   e. both a and c

5. The natural process by which nitrogen gas is converted to usable forms is called
   a. nitrogen fixation.
   b. synthetic nitrification.
   c. the Kjeldahl process.
   d. all of the above
   e. none of the above

6. Humans have added vast amounts of synthetically fixed nitrogen to the atmosphere. When this occurs, _____ increases, but _____ decreases.
   a. used nitrogen; available nitrogen
   b. species diversity; NPP
   c. NPP; species diversity
   d. available nitrogen; NPP
   e. none of the above

7. Carbon dioxide concentrations in the atmosphere have risen dramatically over the past 200 years. The primary cause of this increase is
   a. high rates of natural decay.
   b. human burning of fossil fuels.
   c. agriculture.
   d. all of the above
   e. none of the above

8. The annual increase in carbon dioxide results from two activities: burning fossil fuels accounts for _____ percent of the increase, and deforestation causes the remaining _____ percent.
   a. 30; 70
   b. 40; 60
   c. 50; 50
   d. 75; 25
   e. 90; 10

9. So-called greenhouse gases have been given this name because they

   a. trap the heat radiated from Earth's surface.
   b. are produced by plants.
   c. are necessary for plant growth.
   d. are similar to gases produced in a greenhouse.
   e. none of the above

10. Assertions that global warming is occurring are supported by observations that
    a. many northern plants have increased their growing seasons.
    b. several migratory bird species have shifted their range farther north.
    c. some butterfly species have been found farther north than ever before.
    d. all of the above
    e. none of the above

11. According to ecologists, humans must _____ in order to protect Earth.
    a. note the damage that already has been done
    b. strive to live with the goal of sustainability
    c. reduce population growth
    d. realize the consequences of human activity
    e. all of the above

12. Many species are now _____ because of human impacts on the biosphere.
    a. extant
    b. extinct
    c. in hiding
    d. all of the above
    e. none of the above

13. Match each term with the best description.
    __ land transformation
    __ water transformation
    __ global change
    __ global warming
    __ sustainable living
    a. responsible use of resources by humans
    b. alteration of waterways by humans
    c. changes in land through human activity
    d. sum total impact of humans on Earth
    e. overall increase in Earth's temperature due to effects of greenhouse gases

Conceptual Understanding

1. In the Midwest, a high percentage of land is devoted to agriculture. There also has been an increase in large-scale hog lot operations. A sampling of streams throughout the region shows an increase in _____ due to fertilizer and animal waste.
   a. nitrogen
   b. carbon dioxide
   c. sulfur

d. phosphorus
e. none of the above

2. Vast tracts of rainforest have been clear-cut and burned. This practice has increased atmospheric carbon dioxide levels significantly because the
   a. forest provided a place for the carbon dioxide to be used.
   b. burning of the forest released a large amount of carbon dioxide into the atmosphere.
   c. grasslands that replaced the forests can't utilize as much carbon dioxide.
   d. all of the above
   e. none of the above

3. The major source of excess nitrogen entering the nitrogen cycle is from
   a. bacterial fixation.
   b. animal waste.
   c. synthetic fixation for use in agriculture.
   d. all of the above
   e. none of the above

4. Increases in productivity most likely cause loss of species because
   a. those species that increase in productivity outcompete other species.
   b. the causes of increased productivity in one species do not necessarily have the same effect in all species.
   c. adequate resources are not available for productivity to increase in all species.
   d. all of the above
   e. none of the above

5. Humans disturb chemical cycles by removing large amounts of chemicals from ecosystems. (True or False)

6. Which of the following is a potential result of global warming?
   a. increased ocean levels
   b. destruction of cities and island nations
   c. loss of species
   d. loss of habitats
   e. all of the above

7. Although little on Earth has been untouched by humans, our major environmental impact comes from
   a. land and water transformation.
   b. land transformation only.
   c. water transformation only.
   d. agriculture.
   e. none of the above

8. The draining of wetlands for building residential communities and for use in agriculture is an example of

a. land and water transformation.
b. land transformation only.
c. water transformation only.
d. land conservation.
e. none of the above

9. Human impact does *not* affect climate. (True or False)

10. There is no way for humans to create a sustainable lifestyle on Earth. (True or False)

11. The greatest danger posed by chlorofluorocarbons (CFCs) is that they
    a. are toxic.
    b. decrease the ozone layer.
    c. alter the nitrogen cycle.
    d. cause global warming.
    e. none of the above

12. Nitrogen gas makes up 78 percent of the atmosphere, and thus nitrogen is readily available to plants and animals. (True or False)

RELATED ACTIVITIES

- Write a one-page report describing how emissions standards are used to curb environmental pollution. Try to include statistics.
- Write a one-page proposal for how people of your generation can help move society toward sustainable living.
- Investigate how the United States compares to other countries in terms of environmental impact. Using your findings, write a proposal that could be presented to a local political group.
- Evaluate the evidence for and against global warming. Illustrate this evidence using a Venn diagram for both arguments.

ANSWERS AND EXPLANATIONS

Factual Knowledge

1. b. Humans have modified between one-third and half of all land on Earth. For more information, see Section 25.4, Climate Change.
2. c. Human control of 30–35 percent of NPP leaves just 65–70 percent for all other species on Earth. For more information, see Section 25.1, *Land and water transformation have important consequences.*
3. d. Synthetic chemicals, industrial processes, and agriculture all alter natural chemical cycles. For more information, see Section 25.2, *Bioaccumulation concentrates pollutants up the food chain.*

4. e. Living organisms can make use only of nitrates and ammonium. For more information, see Section 25.3, *Humans use technology to fix nitrogen.*

5. a. The natural process of nitrification is called "nitrogen fixation." For more information, see Section 25.3, *Humans use technology to fix nitrogen.*

6. c. Productivity does not increase uniformly across species, so some species are driven out. For more information, see Section 25.3, *Humans use technology to fix nitrogen.*

7. b. Increases in carbon dioxide in the atmosphere have been caused largely by the burning of fossil fuels. For more information, see Section 25.3, *Atmospheric carbon dioxide levels have risen dramatically.*

8. d. Current estimates suggest that about 75 percent of the carbon dioxide is caused by burning fossil fuels and 25 percent is caused by the reductions in photosynthesis that follow deforestation. For more information, see Section 25.3, *Atmospheric carbon dioxide levels have risen dramatically.*

9. a. A blanket of gases around Earth acts much like the glass of a greenhouse because it holds heat generated by Earth inside the atmosphere. For more information, see Section 25.4, Climate Change.

10. d. Although many politicians and policy makers would like to overlook the evidence, it is mounting along the lines presented here. For more information, see Section 25.4, *Climate change has brought many species to the brink.*

11. e. We are obligated to do all of these things as the bare minimum. For more information, see Section 25.4, *Timely action can avert the worst-case scenarios.*

12. b. Human actions have significant consequences, such as causing the extinction of other organisms. For more information, see Section 25.4, *Climate change will likely have severe consequences.*

13. c. land transformation. For more information, see Section 25.1, Land and Water Transformation.
    b. water transformation. For more information, see Section 25.1, Land and Water Transformation.
    d. global change. For more information, see the chapter opener "Is the Cupboard Bare?"
    e. global warming. For more information, see Section 25.4, *Global temperatures are rising.*
    a. sustainable living. For more information, see the chapter's Biology Matters box, "Toward a Sustainable Society."

Conceptual Understanding

1. a. Fertilizer and animal waste increase nitrogen in ecosystems. For more information, see Section 25.3, *Humans use technology to fix nitrogen.*

2. d. Forests utilize vast quantities of carbon dioxide during photosynthesis. For more information, see Section 25.3, *Atmospheric carbon dioxide levels have risen dramatically.*

3. c. Agricultural practices have a vast impact on the nitrogen cycle. For more information, see Section 25.3, *Humans use technology to fix nitrogen.*

4. d. Resources are limited. An increase in a resource that benefits one species often causes another species (or several species) to suffer. An increase in resources may mean that one species outcompetes another, that resources are not equally available to all species, or that the rise in resources does not have the same effect on all species. For more information, see Section 25.3, *Humans use technology to fix nitrogen.*

5. False. Humans disturb chemical cycles by adding synthetic and naturally occurring chemicals to ecosystems. For more information, see Section 25.3, *Humans use technology to fix nitrogen.*

6. e. Global warming may have diverse effects, including rising sea level and destruction of cities, habitats, and species. For more information, see Table 25.1, Some Consequences of Climate Change.

7. a. Humans have significantly altered land and waterways for their use and have profoundly influenced the environment. For more information, see Section 25.1, Land and Water Transformation.

8. a. This practice has an impact on both land and water. For more information, see Section 25.1, Land and Water Transformation.

9. False. Evidence indicating that humans are influencing long-term climate change is very strong. For more information, see Section 25.3, *Humans use technology to fix nitrogen.*

10. False. Though not easy, achievement of this goal is possible through increases in education, individual action, research, government policy, and pressure from the business world to develop ecologically sound practices. For more information, see the Biology Matters box, "Toward a Sustainable Society."

11. b. CFCs (synthetic chemicals used as coolants) have caused a decrease in the ozone layer. This decrease is serious because the ozone layer shields the planet from harmful ultraviolet light from the sun. For more information, see Section 25.2, *Many pollutants cause changes in the biosphere.*

12. False. Although the atmosphere is composed of 78 percent nitrogen gas, plants and animals cannot use this form of nitrogen. Gaseous nitrogen must be converted to nitrate or ammonium. For more information, see Section 25.3, *Humans use technology to fix nitrogen.*